AND

...

Roc Morin

Book I

Siren Sonata

The history of the world is embedded in the body. Boney projections in the mouth that once anchored fangs. Tails that receded and curled in. Fingered feet that once were hands. And the colors we see - the faintest cast of the solar array. It was all that reached us, all that could penetrate the waters of the ocean above.

The woman was 26. She fell 26 floors. A family on the 14th floor found her right hand on their balcony. The rest of her lay on the pavement below.

Someone said, *Hope she's a lefty.*

I stared at the twin blood trails that led from her head - meandering rivulets revealing the imperfections of the pavement. Miraculously, they ran, like duplicate signatures, all the way to the gutter without touching.

The cops let us onto the roof. It was Pineiro's idea to see where the woman had jumped from. But, there was a wall we had to climb, and all that sooty ductwork we had to crawl through, and finally those sloping glass panels. And, we knew exactly where she had been because the soot was smudged in places, and we knew how - handprint by handprint, the fingers smeared out into long claws - she had slid down that glass and over the edge.

Look, said Pineiro, glancing off. There was a pair of high-heels at the edge, neatly placed, as if by a maid. As if the jumper had expected to return.

Maybe she thought she could fly.

Every detail seemed deliberate, unalterable. And, the more they seemed that way, the more they seemed like decoys.

That's how it was with the Bajan man, 33, his kempt moustache - he who had crawled under his bed to die, or to hide from death. And only his right arm protruding, levitating in rigor mortis, index finger pointing - pointing at the wall, at nothing.

And that Russian who had swallowed poison - how the Lieutenant made the mother promise she wouldn't cause a scene before they let her see the body. How she kept her promise - insisting only, in a whisper, *But he had a haircut yesterday.*

Another call. This time, a hanging.

The cop asked, *Why would she do this?*

And the sister replied, through the mask of her face, *She was misunderstood. Even by herself.*

There was a chair, but she wouldn't sit down.

The chair was no longer a chair.

A Message

My darling, you left your pretty buttons on my bed! Send me your address so I can mail them back A.S.A.P. The thing is, they were left so deliberately perfectly perpendicular to the edge of my bed that it seemed like you meant to leave them there, and even now I'm only thinking you didn't do it on purpose because you spent a good bit of time picking them out and talking about putting them on a suit. But still. You did just forget them, yes?

Cutty

During the hiring process they marched a hundred of us into an auditorium. All day long, we marked little haiku statements true or false.

I would like to be a singer.

I gossip sometimes.

Everything tastes the same.

Everything happens for a reason.

Sometimes my hearing is so good it troubles me.

Nuclear war could be exciting.

The department called it a psychological test, but it was more like an IQ test. It was clear that nuclear war could never be exciting. Nuclear war would be extremely dull. The advice coming down from all the old-hands was: *Be dull.* That, and be prepared for the drawing test.

For the next part of this test, the proctor announced, *you will draw a house, a person, and a tree. You may draw them any way you like. There is no wrong answer.*

Except - draw your windows too big and the house looks surprised. Too small and the house looks angry.

I practiced for a week until I could draw the world's friendliest houses: picket fences, swing sets, smoke curlicues from chimneys. All my people smiled and waved. All my trees were Christmas trees.

In a month, the results came back. I had been classified as sane. Then again, so had Charles Cutty.

My first day on the job, I walked bang into one of Cutty's sermons. His chrome skull-and-crossbones belt-buckle darted and bobbed as he addressed the lounge, describing his favored method for murdering an irksome wife or girlfriend.

Nobody at the station talked to Cutty. They shunned him like the Amish - looked right through him. Even crazy Billy Sand did. And, now too, everyone around was pretending to read a magazine, or shuffle papers, or work the vending machine - but the room was silent. All ears were dialed into Cutty's station, and he knew it.

Now, my skels - ya gotta wait till that bitch is menstruating. Very important. Then ya bring that bitch some roses. Tell that bitch: Hey baby-baby, honey-honey, sweetness. All right. Then whatcha gonna do - ya gonna lay her right down on that bed, and ya gonna proceed to eat that bitch out, ya follow?

He swiveled his head, swirled his eyes around the room. Everybody looked away.

Eat that bitch out?! Surely you must be joking, Mr. Cutty. I wish I was folks, but I am serious as an infarct. Okay, next, you blow, blow air into it - make a seal like a balloon - blow like she's a candle and it's your birthday. Got me? And, pretty soon it's gonna feel like your birthday, cause now that bitch is dead. Air embolism, medics. Untraceable.

He grinned so big, it seemed his teeth would crack.

I remember the day Cutty lost it, truly and finally, Knox recalled later. *Cutty was up front driving. Me and Kamacho had a cardiac patient in the back. Anyway, this guy goes into*

10

arrest. So we're working him up: defib, chest compressions, pushing drugs, everything, yelling at Cutty to go-go-go, lights-and-sirens. But Cutty wasn't having it. I guess he didn't like taking orders. He slowed that bus to a crawl. I mean, mothers pushing strollers were passing us. And Kamacho was - fuck, I thought he was gonna have a heart attack - beating on that little window, calling Cutty every name in the book. And what did Cutty do? Cutty just started singing. I mean, he got on the loud speaker and serenaded the entire fucking neighborhood with a stirring, two mile per hour rendition of Frosty the Mother-Fucking Snowman.

And, you want to know the really fucked up thing about it? I laughed. I started and then Kamacho started. There is this dead guy lying there with a tube down his throat, and me and Kamacho just howling away.

This job will take its toll, you know. My start date was in August, 2001 - so I basically graduated the academy, walked down to the Trade Center, and began picking up heads. You know that children's song: Heads, shoulders, knees, and toes? Yeah. And bones. Human bones, animal bones, and dinosaur bones all mixed together. Really. Animal bones from what people ate, and in one of the towers there had been an exhibit of fossilized dinosaur skeletons. And nobody knew what was what so we sent them all off to the morgue together.

It was so awful down there on the pile, but you wanted to be there, and when you weren't there it was all you ever thought about.

That was about the time the bad dreams started, and the sleep-walking. I terrorized my ex-wife, grabbing at her neck ten times a night to check the pulse, pushing rags at her, telling her to sop up the blood. Half my dreams I was trying to save people, and the other half I had become my patients. I remember one time, I had some kind of mortal wound but I kept reassuring myself: You can control this. Just concentrate on staying alive. People only die because they believe they can. They give up.

But when I have dreams like that, when I wake up in a cold sweat, there is a kind of relief. It means there is a part of me - I think, the best part - that refuses to tolerate the things I have seen. The day I get a good night's sleep - that's when I'll know it's time to quit.

High-Five

Pineiro opened his mouth at the worst possible moment. We were taping this hit-and-run victim to a board. Pineiro was saying something, when the tape snagged, snapped, and flung blood. Pineiro got a mouthful of 14-year-old-girl blood.

He was spitting pink in the hospital sink, gargling nasty hand-sanitizer. Knox had his hand on Pineiro's shoulder - the voice of reason. *She was 14. Probably a virgin. I wouldn't worry about it.*

They called it getting High-Fived, or meeting Henry the IV, or buying a House In Virginia.

Yeah, but they're starting to fuck younger and younger these days, Pineiro choked out. *Even if she was 13, I wouldn't worry so much...*

We'd all had those particularly bad AIDS patients: the walking skeletons, dead-eyed, brain-abscess-zombiefied.

Pineiro had to decide immediately whether or not to start the anti-HIV cocktail. Taking the cocktail meant shivering, sweating, and vomiting in a room for a month. Knox had just come off his month and said the cure was as bad as the disease.

Pineiro's patient, the 14-year-old, was dead. Knox at least got to ask the patient who contaminated him whether he had AIDS.

The patient was adamant, *Shit no, I don't have the AIDS.* Knox was relieved. *I don't have the AIDS now,* the man continued. *I had the AIDS. But, I cured myself. That week I had the AIDS was one of the worst weeks of my life.*

Vera

I don't know what the proper name is for a place like Sunrise Manor. Everybody called it Heaven's Waiting Room. It was a big building where old people went to die.

Vaughn and I gathered our equipment and walked inside. The call had come over as a cardiac arrest. We had both been here many times, and neither of us had ever brought anybody out alive.

When the orderly led us to the room, medics were already working up a ninety-five year old woman. She was on her bed with her clothes cut off, limbs out, head nodding yes in unconscious acknowledgement of the CPR being performed on her. The way the intubation tape pulled up the corners of her mouth, it even looked like she was happy about it.

The man pushing on her chest stopped to wipe his brow with the back of a gloved hand. *Well,* he remarked between breaths, *It's an arrest all right.*

I should hope so, I replied, *otherwise you guys got a lawsuit on your hands.*

Everybody's a comedian, he said. *Why don't you take over for a bit?*

In the late eighteen-hundreds in Paris, a young drowned girl was fished from the Seine. With no marks on the body, suicide was presumed. She was placed with the other unclaimed corpses on public display at the city morgue. The morgue was completely overwhelmed by the crowds that assembled to see her. Though she remained unidentified, there was something so alluring in her enigmatic smile that men broke down and wept.

She was buried finally, but not before a plaster cast had been made of her face. Copies of her death-mask appeared in parlors throughout Europe, becoming the fascination of the age.

A century later, a Norwegian inventor was looking for a face to give his CPR training doll. He found the death-mask. There is a hall at the

EMS academy filled with his dolls, and we all learned how to raise the dead, breathing life into that girl's mouth.

Doing chest compressions on a doll like that, with its weak spring, is a one-handed affair. Not so with real people. By the time I got my hands on her, the woman at Sunrise already had a couple broken ribs, and I felt a few more snap under my fingers. Textbook CPR.

The firefighters arrived next, as back-up, crowding in close, asking, *Are you tired? Are you tired?* They were good guys and wanted to save a life.

The department had just spent a million dollars on some kind of CPR device that measures how hard people push, and the burly firefighters were competing with each other, wailing on the woman to see who could get the highest score.

From the doorway, it looked for all the world like she was trying to fly away, out the window, and their job was just to keep pushing her back down to earth.

All around the room were black and white photographs of people who had meant something to this woman. A lovely girl in a gingham dress eternally kissed the face of a handsome marine. Even with the woman's teeth gone, and the tubes coming out of her, even with her shriveled body and cloudy eyes, it was clear that the girl in the gingham dress had been her.

An orderly arrived to rattle off the woman's medical chart: *Diabetes, Hypertension, Stroke, COPD, Arthritis, Cancer, Alzheimer's-Induced Dementia...* A full catalog of human suffering.

And the final strike, like a verdict being read, *Next of Kin: None.*

What's the most stable heart rhythm? the senior medic quizzed. The answer: *Asystole.* That meant no rhythm, no beat at all. He called it at 22:53.

We pulled the defibrillator patches off the woman's chest. We shoved gauze and torn plastic packaging into a red bag. Everything was clean now.

I took one last look at the photographs. Nestled among them was a sheet of paper in a dollar-store frame. It hung crooked over the bed. It read:

Certificate of Achievement Presented to:

Vera Washington

*For not only being The World's Best Mother
but also My Best Friend*

Signed,

Sunrise Manor

Heartworms

Every evening at a quarter to five, Mr. White's heart stops beating. Fortunately, in these crises, Mr. White always keeps a cool head. He staggers to the payphone, dials 911, informs the operator of his condition, and lies down in front of the homeless shelter to await the ambulance now speeding to save him.

Cardiac arrests are always priority 1, so Mr. White knows he only has to hold on for a few minutes longer. Mr. White is not worried. Mr. White is confident in his abilities. Mr. White has long since mastered the delicate art of living without a heartbeat.

It was me and Vaughn. We missed Mr. White that day but we got Ms. Thorton. The call had come over as a stroke. Having picked up this same woman every day for a week, we recognized the address immediately. Vaughn always attracted the crazies.

I haven't gotten a good call in over a month, he lamented. *I can't swing a dead cat without hitting another nutjob.* To an EMT, a good call meant adrenaline, meant blood.

Listen, Vaughn said. He had that good call sparkle in his eyes. *I'll bet you twenty bucks that after this job we don't see Thorton again for three days.*

I'll take that bet, I said.

There is an old joke where a woman goes to the doctor and complains, *It hurts all over.*

The doctor replies, *Can you be a little more specific?*

So the woman presses her finger to her head and says, *Ouch!* She points out her knee, her cheek, her little toe. It's all the same: *Ouch! Ouch! Ouch!*

The punch line is: the woman has a broken finger.

Vaughn was hunched over listening to our patient - his lips pressed together thoughtfully, nodding respectfully in a flawless imitation of attentiveness. He let her talk about the sand that had seeped under her skin, her orange urine, the mystery pains in her hair. Finally, she fell silent.

Vaughn sipped air, *I hate to tell you my dear, but you've got a case of heartworms.*

The words landed like a blow, but Ms. Thorton recovered quickly. She studied Vaughn's face for a long time. *People can't have heartworms,* she stated. *That's for dogs.*

All eyes on Vaughn. Not even the trace of a grin. *Have you ever heard of anthrax?* he asked her.

Ms. Thorton nodded.

Okay, he continued, *do you know where anthrax comes from?*

It was beginning to seep in now. *Cows?* she offered.

Very good, he said. That was all it took.

The trick is, Vaughn explained, *you have to get to the triage nurse first, before your patient. You present your case: We've got a real EDP here - thinks she has heartworms - claims the heartworms are talking to her, telling her to hurt people. Easy as pie. Oh, she'll be out again. But as you know, the psych ward takes three days...*

And he was right. In three days she was out and calling for more ambulances. She was crazy, but she wasn't crazy enough.

She was like another of our frequent flyers: Dahlia. Sweet Dahlia who claimed to be the wife of God. Wide-eyed Dahlia crying out, *Save me Lord!* as the Roman-faced cops rolled her in a Persian rug with her head poking out. *O, help me Godddddd!* she wailed. *My husband! Destroy their wicked hearts! They are not worthyyyyy! I try to teach them what is right, and they refuseeeee!* And the cops, all six of them, hoisting her down the stairs, shouting *Hallelujah!* like a bunch of Baptist pastors.

Dahlia heard God's voice, so the hospital let her go. They said, *Come back if you hear the Devil.*

On the way out we passed Mr. White. He looked good. He had found his heartbeat again. He lay on a bed with clean sheets, watching TV, wolfing down the five o'clock meal. Nothing crazy about that.

Yes

Jack Rubicon was the prince of the special care ward. He was Hollywood handsome, with a Douglas Fairbanks moustache. He always wore the finest clothes: collared shirts, and ties, and vests.

Neither the moustache nor the clothes were his. It was the nurses who had decided on the moustache - the nurses who spent their own paychecks buying him clothes.

They did ask Jack if he liked it, and in his steady baritone, Jack had answered, *Yes*.

The nurses asked Jack if he thought they were beautiful, and Jack said, *Yes*.

They asked Jack if he loved them, and Jack

said, *Yes.*

They asked Jack if he would marry them, and Jack said, *Yes.*

Yes, was all Jack ever said - all he could say. He had been okay once, but something bad had happened, and whoever he had been, it was all gone now.

Now, he was nobody and at the same time he was everybody, bed-bound and living all these different lives, everybody the nurses thought he was or wanted him to be.

Salt

Christmas Eve. The aftermath of a police chase. A car smoking in somebody's living room. Ornaments and gifts and bricks everywhere. A strand of flashing lights under the tires. The vehicle had ground up the Christmas tree after coming through the wall. That pine scent. The family put on coats and boots and stood by the flickering TV, quietly exhaling steam.

The front of the car was gone, punched into the driver's compartment. Firefighters pulled out the driver, legs intact but dangling like ropes. We carried him off thinking that was the end of it, but the man claimed he had a passenger.

The lieutenant was saying, *Impossible* - that the driver was stoned. The passenger seat was crumpled, but it was clearly empty and it kissed the dashboard. In the back was nothing but a bunch of bags of rock salt. The guy could have been ejected, but where? Flashlights everywhere. People looking up into the trees.

Eventually somebody noticed a piece of shirt sticking out from between the dash and the passenger seat. Like a magic trick, a whole man had been compressed into the leg space. They cut that car to bits getting him out. Firefighter surgeons with pneumatic shears performing a kind of cesarean section, and the man was born again.

He said his name was Wayne. In the back of the ambulance, we cut open what was left of his shirt. There was an old scar on his chest. A cop was leaning in through the open door watching. He pointed to the scar. *That's where I shot this guy last year,* he said.

Hey Gary, said the patient.

Hello Wayne, the cop replied. *You're one lucky bastard, you know that?*

Wayne seemed stumped for a moment. *I'm lucky you're not a better shot,* he said finally.

Gary chuckled. *Merry Christmas, Wayne,* he said.

Merry Christmas, Gary, said Wayne.

The whole chase had started when Wayne and his partner had stolen 22 two pound bags of rock salt. Now one of the cops was asking his superior, *What should we do with the salt?*

I'm not dealing with it, the man replied. *Save one of the bags for evidence and dump the rest.*

So, the first cop began removing the plastic bags of salt one by one, ripping them open, and dumping their contents onto the lawn.

I walked over. *Isn't that going to kill the grass?* I asked.

The cop's crisp words delivered themselves on a plume of white vapor, *Do I look like a gardener to you?*

Motherfucker

The voice of the dispatcher is the loudest voice in our heads. Wherever we hide, it finds us. It calls us by the name it has chosen for us, instantly overwhelming our own feeble inclinations with its steadfast resolve.

The call came over for an injured prisoner. It was some banged-up kid in a cell with a cut on the left side of his head. Vaughn bandaged the kid while I bandaged a wound on the right hand of a cop.

Vaughn was addressing the cop, *Just for the paperwork - what happened?*

He fell, the cop said, the ends of his mouth quivering. *Repeatedly.*

Motherfucker beat the shit out of me is what happened! the kid screamed. *You know I'm underage right? Your ass is gonna get raped the fuck out in jail for this!*
Hey, said the cop rapping on the bars, *just be grateful it's winter. The cold will keep the swelling down.*

Gimme your badge number! the kid shrieked. *Show me your badge!*

Certainly, said the cop, pushing his badge right up against the bars, index finger over the number. *Underage my ass.*

Pull up my pants! the kid demanded. His pants were around his knees and his hands were cuffed behind his back. *They took away my belt. Pull up my pants or give me back my fucking belt!*

Pull them up yourself, the cop said. *Anyway, I thought that was the style.*

It's the style when I fuck your mother and auntie, the kid said. *Show me your fucking badge number!*

The cop fished out his badge again and held it up, backwards. *Did ya get it?*

Motherfucker beat me so bad! I can't even remember my name! I'm gonna pass out!

Go ahead, said the cop. *Pass out. Do us all a favor.*

I'm gonna fuck all your mothers! the kid howled. *This cocksucker beat the shit out of me and none of you assholes even cares!*

We're not here to be your friend, sweetie, Vaughn simpered.

And, that's when the kid sucked air, and spit. Pink saliva running down our faces like honey.

I think you guys left some equipment back in the truck, the cop remarked without turning.

There was a painting, hanging on the wall of that station, right at the entrance. One of the officers had made it. The painting was of the station itself, as seen from the outside - except, surrounding it, where there should have been tenements and bodegas and cracked sidewalks and hustlers - there was nothing but desolation, black and flat.

The station stood alone in that little world of the painting, shadowed by a scene of impending annihilation. The brushstrokes of it licked the edges of a frame unable to contain it - radiant, immaculate: a sky-scraping tidal wave of deepest blue.

Daisies

Destiny was six years old, perched on a park bench, sucking her knobby little knees. There was nothing else the twenty milling cops could do, so on a hot August night, they draped a blanket over her. My partner Vaughn, a bull of a man, made himself her size, kneeling, head bowed, with tucked-in elbows, gently peeling back the edge of the blanket with a single steady finger.

Congested wreaths of jellied tears kaleidoscoped the cop lights. We bent and re-bent, blurred and sharpened, our heads spidering with halos.

Hi, Princess, Vaughn whispered. *Can you walk?*

She didn't blink. Her eyes were open but she didn't try to see.

Do you want me to carry you? he asked. She swallowed. She nodded, hard, and he gathered her all up and reeled her in as if pulling something out of the sea.

In the back of the ambulance, he was magnificent - spinning her onto the stretcher. *Just like dancing*, he sang into her ear - nodding at me to start on her radial fracture, to bandage her abrasions, to irrigate her swollen eye, to wash away the web of blood that had dried running down her legs. All this, while Vaughn kept her distracted in a hundred little ways. He promised her ice cream and toys at the hospital, and flowers.

What kind of flowers do you like? he asked.

I like red ones, she said, *and peach, and blue, and green, and beige, and yellow.*

Well, princess, we'll get you flowers in all those colors, and if we can't find 'em, we'll paint daisies.

She adored him - didn't even wince when I straightened her arm to splint.

Well, I don't have flowers here, he confessed, *but I do have a balloon. It's a blue balloon. I hope you like blue.*

I love blue.

He inflated a rubber glove and tied it off. With the thumb for a nose, he penned a smiling face and presented it to her with a flourish.

We could hear the brothers outside, crying for revenge, the mother screaming, hysterical, sobbing so hard she began to vomit.

After the job, we climbed back into the ambulance, staring off, pretending we were alone. I weighed the tips of my fingers, waiting for a feeling to come and make clear what must be done.

I said, *Sad, what he did to her.*

Vaughn turned, just slightly. Nothing happened.

Listen, he said finally, *I know you're new and all. You might think it's cold, but now that we're out of the E.R., to be perfectly honest with you, I couldn't give a fuck what happens to that family.*

He didn't look over, or smile, or frown. He spoke with the voice of someone just telling you the time.

Since we're gonna be partners for a while, you might as well know how it is. I act sweet because I think it's funny. When the apocalypse comes, I'll be the first to kill and eat another human being.

Tomorrow

I always get a sense of déjà vu in the projects. The same apartment templates repeated into infinity. The patient wore a straw boater. He was a linebacker of a man, but didn't seem to know it. He looked smaller with his palms up, shaking his head, trying to shake off a bad headache. He smiled softly when, in unison, we happened to glance out the window. White clouds over Brooklyn.

He spoke with a voice you had to lean in to hear, *The sky is falling.*

Oh, I hope it's not as bad as all that, I said.

It is, he assured me. *I think that's why I feel so terrible.*

Did a piece of the sky fall on you?

Yes, it did. A black tunnel came and sucked me in.

Pineiro and I glanced at each other. *What kind of tunnel?*

A time tunnel.

Oh, I see.

That's how I found out the world's going to end the day before tomorrow.

The day before tomorrow? That's today.

He nodded somberly. *I know.*

The next morning in Kaiser Hospital's triage, that same man was there, an entourage of cops and medics ranged around. He was handcuffed to a stretcher.

The man was yelling now, *Where's my hat?*

On your head! everybody answered.

Where's my head? he asked.

The whole lobby roared with laughter.

You laugh, he shouted, *but I don't care because I don't exist. You think I'm crazy? You're all crazy because you think you exist!*

Pineiro walked over to him, *What happened yesterday? I thought the world was supposed to end.*

The patient gave Pineiro the kind of look that a father gives his child.

Hate to tell you, but the world did end, he said. *It ended, and no one was saved, and a new world started up in its place. Nobody noticed but me.*

Book II

A Prelude

*W*hen *I was a very little boy,* my grandfather began, *I left the farmhouse and went walking out into the fields. I went far, and when I turned around, the farmhouse had vanished. I didn't know that I had only gone over a hill, and it was hidden behind that hill.*

I looked up and saw only the sky. I looked at all the sky of Saskatchewan. The ground was burned by the sun, and there was only the sound of crickets. Creeep! Creeep! Creeep! Many, many crickets. That suffocating sound! And here I am, a tiny little boy and I'm all by myself. I was alone in this whole world.

Nightshade

I was born without claws, without fangs, without poison venom. My mother and father took care of me. The moon followed me and the wind obeyed me.

Deadly Nightshade with its black jewel berries grew wild around our home. My parents had warned: *Just one berry will kill you.*

We had to find out on our own that rose thorns hurt, because if you didn't know any better, wouldn't it seem just as likely they'd cause pleasure, buried under your skin like borrowed babyteeth?

I remember the nightshade, and that I used to run: just racing everywhere, racing rabbits and deer, bags caught in the wind, bicycles, cars -

escaping through forests of eyelashes, into the shadows of passing airplanes.

My Grandfather, a doctor, had announced gravely, *All that running will enlarge your heart until it bursts.* And when my heart burst, it would shatter glass for miles.

Running at night, feeling that transparency begin to work its way in from the edges. Face getting dark, blurring away. The sensation of a great and terrible hand reaching down from out of the clouds. Only, instead of touching me, it passed right through.

In each thought, I arise anew...

I am there in the thought, and nowhere else...

Anyone can think this thought and become me...

I can be anyone...

After sundown, the windows of the houses became one-way mirrors. From the outside, you could see in, but no one could see out. I saw children at the dinner table. I watched husbands and wives wearing TV-light masks.

He would go upstairs and she would clean up - standing at the sink, her face against the window, staring at her hands. I breathed through my nose so the glass wouldn't fog.

She left the light on and mounted the stairs. Outside was their invitation. A mat that read: *Welcome.*

Under it was a key.

The floors creaked in places and held silent weight in others. I drifted from room to room. Children clutching dolls. Husband and wife sleeping without touching.

Downstairs, if there were still dirty dishes in the sink, I would wash them. It was a test of patience, since I could only use a thread of water. Then, I would open the refrigerator to pile up the remnants of meals the wife had cooked. I ate them with my own forks and knives of quiet plastic.

In the refrigerator, for dessert, there was a little bottle of cherries. I ate them all. Every visit, there was a new little bottle of cherries to finish - until one night, and from then on, there were cherries in a large jar.

The wife had reached out to me. Marshalling the evidence I had left, I believed that she had assembled me in her heart. Even if she only knew my hunger - had mistaken it for someone else's - that was enough.

I searched for that feeling everywhere. Above all, I trusted in honest mistakes to deliver it.

When a child grasped my hand in a crowd, mistaking it for his father's - when a love letter arrived bearing someone else's name - I was so close to being the right person, that for a moment, I was.

Hiroshima, Wisconsin

A lone B-29 Superfortress buzzed the upturned faces of Sunday picnickers. Little girls covered their ears. The bomb-bay doors eased open, and a single invisible atomic warhead tumbled through the air. Then, from the roped-off tarmac beyond: the orange fireball, the thunderous *KABOOM*, the heat flash, and the dirty black mushroom cloud that languidly dispersed off into the concession stands.

The national anthem echoed across the airfield. I was a kid then, with my father at the world's biggest air-show in Oshkosh, Wisconsin. The tarmac was Hiroshima.

After the war, I wandered into a massive hanger filled with sleek vendors, and one rickety booth dedicated to the philosophy of aviation pioneer Alfred Lawson. The elderly Swain Turnbull and his wife Millie waved me over. Millie asked if I liked baseball. She brought out photographs of Lawson from the early 20th century playing in the majors. Swain showed other photographs of Gilded Age tycoons. When I correctly identified John Jacob Astor, he gave me a stack of Lawson's books and took me out to dinner.

Letters from Swain came every week until he died several years later. Lawson taught that the human race is in the process of evolving mind-reading powers. Once we become telepathic, there will be no secrets and no lies. That is how evil will get rooted out of us. We will see it inside a person as clearly as x-rays reveal a tumor.

All of this frightened me. What a terrible feeling to be opened up like that, exposed to everyone - all faults, all shameful desires. I had nodded politely, but it was a decade before I understood.

I was on MDMA, sucking a pacifier in some lonely corner of a party. MDMA: love's slow demolition. Ripples chasing the elastic shore. Love so vast, no action could ever redeem it. Empathy so debilitating, the only possible relief might have been to become a slave.

I stepped out of myself and watched my thoughts arriving like carrier pigeons, one after another: fully-formed messages from a hidden source. I wondered, how would I even know the difference between my thoughts and the thoughts of someone else? Thoughts bear no signature of authorship.

This is what Lawson's telepathy dream would really be like. It would be impossible to tell where one person ended and another began.

When my hand, pricked by a thorn, causes you the same pain as me, you will pull my hand away, and then it is our hand.

Santiago

A letter from the wilderness. An invitation. No mention of the stroke. Father John was in Guatemala ministering to a jungle-bound archipelago of 80 villages. He arrived to nuns scrubbing blood from the walls and ceilings of the new mission house - the deposed dictator's former prison. John slept in the interrogation chamber.

At the airport in Guatemala, nobody met me - the first sign that something was wrong. The second sign came when I took a taxi to a walled compound in the capital's slum. It was the address John had given, but none of the priests there knew of a man by his name. I showed them John's picture. The priests identified the man in the picture as Santiago.

After several calls, the truth emerged. John had renamed himself to better accommodate Latin tongues. I had been asking for a man who no longer existed.

Santiago had forgotten I was coming. He had forgotten the stroke and all that came after. He had forgotten how it had damaged his memory - damaged it so badly he had to ask again every morning why I was there.

But his charm endured. He embraced his loss, even believing it had brought him closer to God. As God is everything, God can only create through subtraction. Ignorance was God's great invention, and with it he made the world.

The Bible states: In the beginning there was nothing. What it really means: In the beginning there was everything.

Whenever Santiago saw a flock of birds or a grazing ox, he would instruct, *Watch them pray. Animals pray constantly. They lived with us in the Garden of Eden, but unlike man, they never ate the apple.*

Ultimately, he turned against memory itself. His joke about memory: *Isn't that the definition of insanity, seeing something that isn't there?*

These things he spoke - I remember them so vividly because he repeated them so often -

every day, and always as if for the first time.
And I never let on, always trying to listen in the
same spirit - the spirit of the first man. But
what an unexpected gift - another chance. I
made myself better for him. I learned what
responses curled his lips, what jokes creased
his eyes. I even spoke his words back to him as
my own.

We had the same conversation every day, but
every day better, more refined. We strolled
from landmark to landmark in the landscape of
his ideal memories - where the purpose of
every pause and every digression was only to
provide us the opportunity to forget, to
estrange us from the things most beloved.
Places only departed from in order to return -
and the time spent away, to sweeten the
homecoming only.

La Rumorosa

The mission was based in the town of La Rumorosa, a jungle outpost. The townsfolk and the jungle were at war - an obscene conflict between shapes and patterns. At night, gathered around our light bulbs - the sounds in the distance: infinite teeth and claws scratching away, obliterating the road and all we had made. Vines creeping into it, trees thrown down over it, storms eroding it, rockslides obscuring it, rivers stealing it away. Every day, with chainsaws, ropes, and shovels, we scraped off the scab.

Santiago and the roosters crowed together. Untangling mosquito nets, we sat for a simple breakfast with Father Esteban and the fresh-faced boys from the seminary. Everyone chatted happily and made jokes without malice, jokes about things like having a large appetite or mistaking sugar for salt. After breakfast, it was Santiago and I in his black pickup truck.

Teenagers had claimed the edge of town, kicking the ground and throwing invisible stones at the dogs. The boys would pinch the girls and whisper about a certain insect with a poison bite. The victim dies within the hour. The only cure for this bite is sex. A few minutes drive and we had crossed their lines.

We drove for hours, slow as a funeral procession. The road was barely a road. When we finally reached a remote village, all voices shouted, all bells clanged. The villages had no electricity: no TV, no telephones. It was a spontaneous holiday, our arrival and the mass Santiago would perform for them - a mass of three or four hours. A year's worth of babies to baptize, lovers to marry, sick to anoint. A year of sin to be confessed and have washed away.

In the most obscure lands where the roads did not go, places we had to hike to, the natives spoke only in arcane dialects. I returned from these excursions sporting little bald spots. Ten hands touching my head. First tousling, then light tugging, and finally cutting out tufts of my blond hair. I knelt to make it easier.

I studied their feet. They had beautiful feet with straight toes not touching one another but feeling the earth. All their feet were like this because they never wore shoes. They smiled the kind of smiles that had never been practiced in mirrors.

On this trip, we brought supplies for them. We had many nice things. We had pictures of a man nailed to a cross. We had shoes and we had mirrors.

Thirst

A buzzer sounded back at the mission house
back in La Rumorosa. Santiago unlatched the
groaning door. Through the jungle, in the rain,
five Guatemalans had walked for eight hours.
One man carried a box by forehead sling.
Inside was his dead son, several days old.

The stench of decay from that box: a fragrance
like a favored flower that the fields let grow too
sweet. Sweetness upon sweetness into
repulsion. Vitality to the point of cancer - like a
skin so healthy it grows between the fingers
and over the mouth.

The Roman Catholic Church teaches that all
are born in a state of disgrace. *Behold, I was
shapen in iniquity; and in sin did my mother
conceive me.* The gates of heaven are barred to

the unbaptized. The father was begging Santiago to open the gates for his son.

Santiago gently lifted the infant to pour holy water over his brow. The boy's mouth was frozen into a zero. He looked so thirsty that I willed the water into his mouth, but it only slid around his lips and drummed into a metal basin where it sat for the night. By the following morning, the basin was dry.

If I did not know what wind was, I might believe it was the hand of God running fingers through my hair. And if I did not know about evaporation, I might have, with a luminous mind, observed a flock of ravenous infants all perched at the brim of that baptismal basin, eyes rolled back in their heads, sucking up every last drop.

John

Seven years later, Santiago was John again, back in Saskatchewan at a care facility.

He doesn't even recognize his brothers, the staff said. *He used to chatter all day about hidden rooms without doors. He used to cry, Fire! when we switched on the light above his bed. He used to mutter about the Devil stealing his slippers. Well, he doesn't say anything now. He's dying.*

I delivered a newborn once. He was pink, but he wouldn't nurse. I stroked his cheek, the way they taught us, to activate the rooting reflex. Eyes sealed, the infant's lips arced to seek the breast, to drink. All infants do this. Departing infancy, the reflex submerges, displaced by the powers of volition.

There was John, the healer, all balled up on his side with the orderly adjusting his nasal canula, his oxygen supply. And the orderly, just in passing, happened to touch John's face. Just a touch was all, before John pushed his chin into the touch, his arms crawling up his chest.

There was no woman to reach down to him. There was nothing to be offered. His mother had been dead for twenty years. Still, his mouth found the rhythm: open and closed, open and closed. In that room, on the last day of his life, he sought his mother's breast.

Book III

Sangin

A confidential morning brief: *The Taliban has obtained 21 ice cream carts which they intend to pack with explosives and detonate in downtown Kabul.*

Piles of trash explode. Roads explode. Cars explode. Children explode. Donkeys explode. The soldiers call them DBIEDs: Donkey-borne improvised explosive devices.

We were above all that now. The back hatch of our bird gaped like a big hungry mouth - its tethered machine-gunner leaning out over the void, scanning for muzzle flashes as we dove at the landing zone. Rockaby evasive maneuvers. That cyclone of yellow canister-smoke. The tail-gunners got so accustomed to it all, that once we were high enough, you'd often see them dozing off with a leg kicked out rakishly over the edge.

Sangin is the front line, Colonel Fitzpatrick had explained back at Leatherneck. *Of course,* he added cryptically, *an insurgency has no front lines.*

Airborne, we zeroed in on the confluence of the Helmand and Musa Qala rivers - the world's most productive poppy-lands. Lush expanses of red, white, and purple. NATO had ceased their defoliation programs years ago.

As our engines choked off, the Marines hustled out of bunkers and into the open. Their outstretched palms reached us first for a quick exchange of crushing handshakes. We trudged off to Forward Operating Base Nolay on a hilltop overlooking Sangin city. Only a few years ago, Nolay was packed with 6,000 coalition troops. Now there were 60.

It was the same story everywhere. Bases across the country were being abandoned - returning to desert. There were vast scrapyards of machinery - million dollar MRI machines waiting to be chewed into a hundred dollars of scrap. *There's not an Afghan in the country with enough technical know-how to operate this thing,* a soldier noted. There were roads and there were gates to nowhere.

The room we are speaking in now, a Captain informed me, *will have gone up in flames by this time tomorrow. We can't give it to the Afghans because the plywood's chemically* treated. *If they used it in their cooking fires, it would poison them.* There was an uninterrupted column of black smoke in the distance. That smoke was the base.

The brass said things like: *Every closed base is a victory.* It was the companion phrase to the other slogan of the day: *The Afghans are in the lead.* They even have an air force, NATO boasted. We gathered to watch an ANA helicopter swoop in, pick up fake casualties from a fake battle, pose for pictures, and then soar off. *Now, that's what victory looks like!* a Major shouted over the roar. He mentioned later that ANA helicopters are restricted to limited flights through friendly airspace because a crash would be catastrophic for the country's morale.

By the time I got to Sangin, the Marines were operating out of a small corner compound sealed off by a mud wall from 4,000 ANA.

A properly-maintained mud wall is surprisingly strong, Captain Naughton stated. *I saw a wall like this stop an RPG.* But, ever since the place had been seized from a drug lord five years ago, the fortifications had not been maintained. None of the occupants were familiar with Afghan mud-building technology. The walls were collapsing, and had to be buttressed by prefabricated concrete slabs.

We dropped our packs in a tent before meeting up with Naughton, an interpreter, and several other grunts for a tour of the ANA sector. It was December, and we passed a little plastic Christmas tree. Instead of ornaments, it dangled empty tobacco tins of red and green.

Naughton opened the padlocked gate dividing the two sections. It was purposefully made to look minimally-secured. *We don't want the Afghans to think we don't trust them,* he noted.

As we passed a soccer field on the way to the perimeter, I asked if the Marines ever played matches with their Afghan counterparts. *There have been green-on-blue attacks over things like that,* Naughton replied. *We'd stand a good chance of winning, so we don't play.*

The Captain strolled around like the mayor of Nolay, with jokes and salutations for everybody. We paused in the sunlight to chat up an Afghan infantryman. Everyone laughed as Naughton and the blue-eyed Afghan playfully squabbled over who looked more like Alexander the Great. The round snapped just over our heads - not the cordial *POP* of distant fire - but the rude *CRACK* of a near-miss.

That was close! Naughton shouted into the side of my helmet. Everything sounded far away. We had jumped into a bunker of sandbags and loosely-fitted slabs of bullet-proof glass. They had been made nearly opaque from the cobwebbing of intercepted lead. Our grins glowed white in the dusty dim as we crouched like giants under the low ceiling. Abdul, the sentry on duty, grinned back, standing unbent with his M-16. Everyone exchanged *Salaam Alaikums* and shook sweaty hands.

Abdul was too young to even grow a beard, but he had been fighting long enough to have developed a soldier's version of perfect pitch. He could tell what guns were firing and from where by sound alone. *PKM,* he stated - a Russian model. With his finger, he traced back the trajectory of the round to a hill about a klick away. A sentry later mentioned that he had seen our shooter through binoculars peeking out from behind a building. A little girl was playing out front - placed there, the Marine claimed, as a human shield. Her presence ensured that fire could not be returned.

We stood behind the sandbags for a while listening to AK-47s pop in the distance followed by the canned thunder of RPGs. *It's good to hear the boom,* Abdul quipped with a laugh, *Refreshes the mind.*

That night at Nolay, everybody's mind was refreshed. I laid half-asleep in my rack, under a stone-age sky of infinite stars, listening to the cadence of the gunfire - its mutating rhythms. As the sound merged with my dreams, sometimes it became the insistent knocking of an unwanted guest, sometimes a galloping horse, or ripping fabric. Sometimes it was a skipping record - a song that couldn't quite get started.

Just a week prior, an interpreter had half his skull blown off by a round while he slept in this tent. *I think about that every night when I go to sleep,* one Marine confessed. *Will I ever wake up again?*

Sleep on your side, another man joked. *Makes a smaller target.*

The Taliban prefer night fighting because they can change positions at will, while the ANA remain attached to their guard posts and road blocks. The sentries had nothing more than flashlights to shine into the blackness. *These guys need night-vision goggles,* Naughton insisted, *The Taliban fix their rifles on the Afghans' positions during the day, and then wait for the flashlights to go on to adjust their aim. It takes five seconds.*

The ANA used to have night-vision goggles, a Marine accountant later confided, *but they sold them to the same insurgents that they're fighting against.*

I bundled up and wandered out of my tent during the gunfight's crescendo, doing a circuit of the sleeping Marine complex. The stillness demonstrated how much things had changed.

A couple of years ago, it would have been the Marines doing the fighting. Now, it was the ANA's battle. The only men awake inside the mud walls were the sentries and a handful of technicians monitoring the action from overhead. They had an unarmed drone that roared like an industrial vacuum cleaner.

Whenever we send this up, the controller explained, *it sends the Taliban running. They probably think it shoots laser beams.*

The next morning, I awoke to the sound of Sargeant Amaker's big lungs belting out old drill instructor hymns just for the joy of it. His voice was an instrument of terror and he loved it. *I don't know but I've been told: Fuck you Corps, now gimme some more! Eennie-meenie-meiney-mo-mo!*

As I washed up in the sink, the water ran beige off my face with a night's worth of dust. I swallowed a cup of black coffee sent by some Nebraska school children on my way to the ANA's Battle Update Brief. The officers met for a PowerPoint presentation in the basement of a captured mansion. The numbers were unclear, but several soldiers and policemen had been killed by gunfire and IEDs in a series of six Taliban ambushes. Insurgent casualties were unknown, but presumed to be high.

Afghan General Zamarai sat with folded hands behind a desk that seemed like it had been dragged through the mud. The last word was his. He stood and congratulated his men on a successful battle, and then harangued them for the squandering of ammunition. *The soldiers get scared and fire without seeing the enemy! For one Kandak to fire eighty-three 82mm mortars is wasteful!*

When I requested a detailed account of the previous night's fighting from the Marines, spokesperson Major Paul Tremblay responded. *From our perspective now as advisors rather than participating directly in the combat, it can be difficult to piece together the specifics of what happens during any given firefight. What I can say, however, is that the sporadic ineffective fire that came in that day was met with accurate, overwhelming fire from the 2nd Kandak soldiers and Recon Tolai soldiers respectively.*

When I asked to go out with the Afghans though, everybody called me crazy. *Do you like your legs?* one Lieutenant asked. *The ANA are out there in fucking unarmored pick-up trucks.*

On my last day in Nolay, we piled into MRAPs - dinosaur-sized armored vehicles. They drove us to an Afghan artillery emplacement atop the adjacent Heran Hill. Orchestrating the encounter was Captain Dewson, a soft-spoken man with a broad, hair-trigger smile. I had variously heard him described as the Justin

Beiber, the Jay-Z, and the Keith Richards of Nolay, because the ANA troops supposedly loved him so much. He proudly wore a bullet medallion around his neck that one of his Afghan trainees had made for him.

When we clambered out of our vehicles at the summit, it wasn't just Dewson who got the celebrity treatment. The Afghans flocked around us all - shaking hands and patting backs. They handed us cellophane wrapped cakes with tea and offered to slaughter a goat for dinner. They gleefully proposed arm wrestling matches and dance competitions. The Marines shook their heads warily and smiled. The artillerymen settled for having their pictures taken with us in a hundred different combinations.

I respect these men a great deal, Dewson confided as my shutter snapped away. *These are the toughest men in Afghanistan, and you can see how proud they are about it. Look at their uniforms. They live in these squalid conditions, but their uniforms are always clean.*

That night the Osprey came for us again. In the darkness, it grew - a black menacing shape that could only be seen by the stars it blotted out. The roar was apocalyptic. I shouted my goodbyes. Nolay's commander, the fatherly Colonel Douglas, was there to see us off. I thanked him and wished him well. *Stay safe, sir!* I yelled into his ear. A smile broke through Douglas's perpetual expression of benevolent concern. *It's not about staying safe!* he bellowed. *It's about the mission!*

With that, we lifted off from Nolay, heading back to the massive city-sized Camp Leatherneck, 20 minutes away. Our hosts had confidently called it, *The safest place on earth.*

Strapped in, I gazed back through the Osprey's open hatch down into the Sangin night. The horizon rocked back and forth as I watched the muzzle flashes of the insurgents and the tracers of the ANA below playing out their endless call and response - mysterious signs of life in the void. The thought was tinged with a kind of tenderness: *They need each other.*

That night I bunked down with a logistics man. When I arrived, he was tying cardboard cutouts of his hands to a string the length of his arm-span. He planned to mail the thing back to his wife in Atlanta, along with instructions that she should hug herself with it.

Slavyansk

They had blown the bridge at Semenovka two weeks ago, but there were still no signs. The road climbed steadily to the crest of the bridge, and then suddenly, there was no bridge. The little white Lada had been carrying a family of four when it went over. Most cars plunged into the Bilenka River below, but the driver of the Lada had been in a hurry. We discovered the wreck wheels-up on the opposite bank - leaking egg yolks, blood, and a gasoline rainbow.

Looking up, I could see another family perched delicately above the void - tiny figures inhaling their own car's glorious burning tire smell - the smell of survival. They peered down at what might have been - as a lone scavenger picked through the wreckage, searching for unbroken eggs.

It was local businessman Ilya Lazarenko who sent his crane to drag away the wreck. The act of removing one smashed car from a smashed landscape - to Ilya, it was progress. It brought his village just a little bit closer to the way it had been.

He did this, even though he is convinced that the fighting will return. *Absolutely it will*, he insisted wearily. An official within the Donetsk People's Republic revealed to him in advance that the retreat of the rebel army from adjacent Slavyansk was purely tactical: the DPR is massing their forces for a strike against Odessa. This was the same official who had predicted the destruction of Ilya's home at the crossroads of Semenovka. His house had been the large one with the red roof and the two cats - one piebald, the other black.

Ilya's wife, Nastya, had foreseen the destruction even earlier - in nightmares. They started last November, as protesters gathered on the Maidan.

Nobody was thinking about war then, she added.

Still, the visions came. She saw herself and her husband crouching and hiding from gunfire in the ruins of their home. She saw their betrayal, and their execution.

Maybe it means that will happen too, she contemplated. *I don't know. I never believed in dreams before this.*

The official news is not reliable, Ilya muttered grimly when asked about the status of the war.

There was that story the Ukrainians told about the DPR crucifying a boy in nearby Lenin Square. I heard it the first time in a Kiev night club. In Odessa, school teacher Iryna Pietrova recited her version, adding that every time she hears it, the boy gets younger and younger.

He was three years old, the last I heard, she added. *By next year, I expect he'll be a newborn.*

Meanwhile, DPR radio bulletins announced that Ukraine was printing new maps of the country without the name *Slavyansk* on them. The implication was that the Ukrainian army intended to wipe the city off the face of the earth.

The sunflower fields were at our backs. We arrived at the edge of Slavyansk to the clipped thunder of artillery fire. One soldier suggested that it was troops clearing rebels out of a nearby forest. His friend assured us that it was the shelling of Artemovsk. A third soldier speculated that an artilleryman had gotten drunk or become insane from the war and was firing at nothing.

Our Lada shook as we raced through the city center over tank-tread-dented concrete. It was here, on the outskirts where the damage was greatest. The Topopolyok school for special-needs children was in ruins, as were many of the adjacent houses.

Aleksandr and his wife Lena were shoveling rubble from their gutted home. The only possession spared by the inferno was a ceramic lawn gnome.

This used to be a two-story, Aleksandr said, shaking his head. The upper floor had disintegrated.

It was the Ukrainian army, Lena claimed. *I don't know what they were aiming at. There was a rumor that rebels were in the school. It was only a rumor. They were actually 200 meters away. Both sides were firing carelessly.*

It was the DPR who did this, another man insisted.

Most people here don't care which side wins, a woman added. *We just want the shelling to stop.*

In Semenovka, residents stepped over unexploded rockets embedded in the ground, as an adjacent minefield exhaled the putrid stench of corpses whenever the wind shifted.

Boney dogs drank rainwater out of shell-craters as they waited for their owners to return.

A 60 year-old man talked about hiding in his basement during the battle to guard his chickens. He accused the Ukrainian army of using illegal phosphorous bombs.

They made a fire that we could not put out with water, he lamented. *It just burned and burned.*

He claimed his name was *Aleksi* - but a faded tattoo across his knuckles spelled *Ivan*.

We dodged craters on the road to Nikolaevka, near the front lines. An apartment block had been eviscerated, forming a kind of jagged, open-air courtyard with the innards of each apartment exposed to the sky. Standing on the road I could see into a babushka's second-floor living room: her couch, her bookshelf, her decorative plants all huddled away from the precipice. She was sobbing as she swept clouds of dust over the edge, through where a wall used to be. She had been sweeping the same floor for days. The only thing she wouldn't clean was the blood spatter beside the radiator. It was all that remained of her daughter.

Her third-floor neighbor Sergey had been lying in bed when the explosion occurred. His mother, his girlfriend Oxana, and her friend were cooking in the kitchen. The ceiling came down on them all. Sergey and Oxana were trapped under bricks - calling out to one another until a Ukrainian rescue squad extricated them half an hour later. Sergey's mother and Oxana's friend had been crushed immediately. Oxana refused to ever return, so Sergey was alone, slowly sifting through an album of old photographs found in the wreckage. It took him a minute to realize that they belonged to his mother. He had never seen them before.

Do you think this was gas? he asked irately, gesturing at the emptiness around him. The remark referred to state news reports claiming that the explosion here was caused by a broken gas-line.

It was an airstrike, he insisted, implicating the Ukrainians.

One by one, the residents opened their fists to display the fragments of shrapnel they had found at the scene.

Maxim beckoned us to the rear of the building. He grinned as he hoisted up a heavy segment of what they claimed was part of a rocket.

He simpered as he asked it, *Does this look like a gas pipe to you?*

Civilization

Cryptic advertisements in every Chinese city promise:

A symbol of urban civilization

COLLECTION, RESPECT AND DIALOGUE

An oriental era biography facing to the world

Millenium Mansion for Aristocrat

THE WORLD'S COVE

LOOKING BACK, RETENTION AND REVITALIZATION

Better city, Better future

The inheritance-exceeding

Court through Times Peaceful World of Honor

SOPHISTICATED, REMOTE AND ETERNAL

The delicate life

These slogans, these advertisements, camouflaged the walls being built - the new walls rising together throughout the nation, encircling the living neighborhoods of old stone-gate houses and mass-produced concrete apartment blocks. One day the walls were not there and then they were and nobody could guess where they would appear next.

The meaning of the walls is that everything inside will be destroyed. But not right away, not all at once. There is still time for ivy to grow, for birds to build nests, for paint to peel, for graffiti, for dead dogs to turn to bones, for kids to throw stones, for excrement to bleach white in the sun.

Most of the homes empty quickly. Scavengers remove the windows and doors and the wind roars through all rooms without slowing. The rooms contain forgotten objects: the broken accordion, the three-legged chair, the crumpled clothing, the red lace bra, the child's toys, the bamboo cane, the playing card, the dish, the birdcage, the strings and the pieces of things - unrecognizable, with their meanings all carried away.

The buildings are changed into piles of rubble one by one. Not in any methodical way. A house here, a house there. One day a roof disappears and then nothing happens for weeks. A single large hole is gouged in the side of an apartment block and a year passes. Mountains of rubble rise, fall, and shift position from day to day. *Was I here before?*

Armies of scavengers impose order - at first, in the simplest possible way: brick beside brick, tile beside tile, glass beside glass - each in its own pile, like beside like.

For blocks in all directions, the land was flat as a mahjong board now that the rubble of one hundred houses had been scraped away. They hauled it to the eastern plains, I was told, to make mountains for ski resorts.

Just one house remained, dead center, a Qing Dynasty Era stone manor. Voices flew from its windows. Laundry hung from its gate. At night, all lights blazed. Why was it left there standing alone, untouched, before the glacial wall of encroaching towers, all rising as fast as iron can be ripped from the earth? It is because the demolition workers were quartered inside, and on the last day, after breakfast, they exited, turned around, and tore it down too. And, from habit, the last man out closed the door.

As I walked through these places with a camera, residents - hold-outs against the future

– stared at me. When I was alone, a few men and women passed me documents: papers folded down excessively to hide in a hand. They scampered off, not looking back, like people caught in the rain.

Two overseers from the adjacent construction site stormed over, flailing arms, shouting. They wore knock-off U.S. Army uniforms. They shoved me and snatched at my camera.

I wandered onto a street where a crowd was gathered. Half the crowd was police. Everywhere, shattered glass sparkled. Men sat on the pavement, refusing to stand. An ambulance driver explained in broken English that there had been a riot.

In the city of Taiyuan, a man led me to the broken foundation of what once had been a house. He made gestures, rebuilding the house in the air with his hands. He pointed to himself, again and again, until I understood that this had been his house.

He brought me to his bicycle repair shop around the corner. His wife made tea. The man laid his photographs on a table, one at a time:

A smashed window. A door covered in excrement. A house on fire. Crowds with banners protesting in front of government buildings. A bullet hole in glass. Bulldozers and backhoes. Sheet-metal barricades. Walls crashing down.

Taiyuan

Documents From Mr. Li:

A war is raging in Taiyuan...

...On July 18th without warning, the 7000 families of Longtime District were issued notices to abandon their homes. The compensation offered was the same price per square meter for everyone, constituting between one-half to one-seventh of market value. Three days later, the evictions started...

...Early in the morning, two cranes began to advance on the Lam family home. The cranes were accompanied by men as plentiful as a herd of goats, all shouting. In the front ranks they wore black-shirts and army helmets and carried clubs. Behind them were the men with *the red and yellow patches on their arms, the officials. In the rear were policemen from several different agencies. As they closed in on the Lam family home, the residents gathered, throwing flaming torches and igniting propane tanks. One of the cranes was consumed by flames. The aggressors retreated...*

...They then advanced on the home of the four Kin brothers and dragged them all away. One of the wives was so distraught she began to vomit blood...

...The way they behaved, trafficking with corrupt officials, taking our golden land, is it any different than robbery? We are asking the government, what did we ever do to deserve this treatment?

Khilagay

We twisted up into the sky beneath a blur of rotors - mum in the numb motor din. The shadow of our helicopter flit below, yawing in the undulations of the sorrel earth, hurtling peaks, and hurling into dark valleys.

My military escort Lieutenant Cartagena and I had taken off from Mazar-i-Sharif, a city built 900 years ago in desolation - the site revealed in a prophetic dream. From the scattered earthen villages that passed beneath our craft to the army outpost we sought, every settlement seemed tinged with that same fantastic quality - accidents of life in a dead land.

We reunited with our shadow on the airstrip of a base at Khilagy. I jumped onto the tarmac, squinting through the heat blast. The American soldiers of Blackfoot Troop bounded out to meet us. They were tall and strong, engendering visions of Thanksgiving dinners in places like Battle Creek, Michigan, and Louisville, Kentucky.

There was Sergeant King, Lieutenant Morgan, and their corpsman Specialist Singer. They welcomed us by asking our blood types - something practical and still somehow strangely intimate. They hustled us away, out of sight of the mountains, through the gauntlet of concrete T-walls and gravel-filled HESCO barriers that formed the outpost.

We received our briefing in the barracks, leered at by a grinning cartoon skeleton drawn onto the wall in black magic marker. King gestured to a crucified map beside it. *This is us, sir,* he stated, pointing to a dot near the bottom. *And this,* he continued, sweeping his hand over nearly a thousand square kilometers, *is our platoon's area of operations.* He pointed again to a mark about five klicks away. *Just to give you an idea: it takes us 45 minutes to get here.*

The upper third of the map encompassed a slice of the Kandahari Belt, a mountainous region still occupied by the Taliban. *They have so much freedom to maneuver around there,* King explained, *because it's very hard for us to get in unless we're on birds. They're definitely skilled mountain climbers compared to us. I*

86

mean, they've been doing it for years, growing up in it.

Have you tried to clear them out of there? I asked.

We kind of have the mindset: we don't screw with you, you don't screw with us.

Do they feel the same way?

We don't know, King shrugged as the room rang with laughter. *We just do what we have to do. Get up there and get out. So far it's been good. Knock on wood.*

Morgan rapped his knuckles on the table.

So, I continued turning back to the map, *What's the line of demarcation between the Taliban and you?*

King shook his head, *We don't know.*

Suit up, came the order, *the Afghans are ready for us.* We strapped into our bulletproof vests and helmets. The men hoisted M-16s. The 30-pound vest embraced me like a beautiful woman.

We strode through the rest of the camp on our way to the gate. The place was in the process of a controlled abandonment as part of NATO's scheduled withdrawal. Half the buildings were empty, already beginning to fill with the ubiquitous red dust that coated everything. I accepted a cigarette from King, saying that at least it would filter out some of the dust.

Blackfoot Troop would be the last to leave their bootprints here in early 2014. They would pack up whatever the army still wanted and surrender the rest to the Afghans they had trained. That scene had already played out a year ago at the prison we were now headed to, on foot, across a lonely stretch of sand.

Beyond the wire, the soldiers fanned out, rifle snouts rooting. We moved as a herd into the hot mirage-land of the heat shimmer.

The silence was as full of tricks. Silence as premonition. Silence, accompanied by the feeling that for just a moment, between each and every heartbeat, we begin to die.

The Afghans met us at the prison gate in grey fatigues. There were handshakes all around, with palms over hearts in a gesture to convey sincerity. I seemed to be the only one who noticed in the midst of our loud maneuvers, the mute shape slipping through and diminishing into the desert over the shoulders of our hosts. The figure was cloaked in amorphous blue fabric and without a face. Only the wind, here and there pulling taut the chadri garment, flashed rude glimpses of the womanly body beneath. *She must be a visiting wife,* someone remarked.

We were escorted to a room in the administrative building to wait for the warden, Colonel Yaya. The Americans jawed on couches, slumping among slouching piles of shed rifles and vests, their weapons always within reach.

The troops, Morgan noted, were in a delicate position with their Afghan counterparts. Remaining well-armed in their presence would be an insult - a sign of distrust. But, laying aside weapons would leave them vulnerable to the kind of insider attacks that have surged in recent years. Word had already reached us of an Afghan Sergeant gunning down three of his trainers in a neighboring province just that week. Each unit had to find a compromise. For Blackfoot Troop, that meant leaving one man in the corridor with his finger resting just above the trigger of his M-16.

The Taliban is extremely smart, King added. *They're very latent recipient. They'll wait and wait and wait for years before they do anything. One of the high Taliban leaders could actually have a job inside this prison. If he's not called to do something, he's going to act like a normal person. But, when they get that phone call, they'll immediately turn around and do what they're supposed to do.*

Just then, Colonel Yaya entered, briskly fingering the prayer beads that never left his hand: click-click-click. His face was weathered like the landscape. He waved us into his office where framed photographs of president Hamid Karzai and warlord-martyr Ahmad Shah Massoud hung. These two figures are so revered in the north that their pictures affixed to windshields are often used in lieu of license plates. They prove that the driver has the right convictions.

As I settled into the Colonel's leather couch, Morgan leaned over. *Just a few things,* he whispered, *Don't cross your legs, don't show the soles of your feet, don't ask about his wife, and make sure to drink the tea.*

Yaya himself began via translator in the traditional way, thanking us effusively for honoring him with our presence. I responded in kind. Business proceeds slowly, after long pleasantries. I asked about his tenure at the compound.

The Colonel was brought in a year ago to replace his predecessor following a prison break. Yaya was now responsible for 650 inmates divided into three isolated populations. Around 500 were common criminals: murderers, rapists, and thieves. 19 were women, housed in another enclosure veiled by taut floral sheets. Lastly, there were about 100 insurgents - *mostly failed suicide bombers and IED planters*, the warden stated.

Those 100 men were the reason the Colonel's telephone was always ringing - dark voices promising death for him and his family. Yaya spoke of a recurring dream in which the convicts had escaped and were after him. He would bolt out of bed in the middle of the night, dial the prison, and insist on an immediate head count. *My one year as warden has felt like 30*, he mused with a delicate smile.

On our tour of the inner sanctum, Blackfoot Troop were asked to disarm, piling high their weapons on a flimsy folding table. The Colonel led us to the empty prison school, where a shiny new book lay on every desk. He led us to a modest manufacturing facility where several inmates were weaving and tailoring shirts.

He led us to cell block four. Constructed over a sinkhole, the building was shot through with fissures and on the verge of collapse. Taliban prisoners had scrawled what my interpreter called *love poems* all over it. I pointed to one and asked what it meant. He shook his head. *Impossible*, he insisted.

We were led to the insurgent wing. It was just an open yard surrounded by a chain-link fence. The convicts were huddled in a far corner, crowded into the diminishing sliver of shade cast by the barracks. Each side stood facing the other.

Eventually, one prisoner stepped forward, then another, gingerly at first, hesitating periodically, half-curious, half-wary. The boldest one, in a pakul - burly and bearded, reached out to touch the fence, lacing his fingers in it. He fixed me with lapis-blue eyes. The prisoner touched fingers with an Afghan guard, sharing a few jaunty words. Soon they were laughing so hard you could see the red of their throats.

Yaya raised his hand. Facing the prisoners, he spoke with a halting cadence. The inmates quickly eyed their feet. When he finished, he turned on his heels, cutting twin arcs in the dust. I watched him retreat down the wire corridor, trailed by his gamboling guards.

What did he say? I asked the translator.

He say, Don't kill anyone. The Qur'an say make peace. In our religion, if man kill man, they will go to the Devil.

At my request, the translator questioned the prisoners. What did they think about the Colonel's words?

Only their eyelids moved, flickering away the dust.

Convinced they would not answer, I turned away. And that's when I felt it: a sharp sting on the back of my neck. *Fuck!* yawped Singer, grabbing his arm. I spun to see the prisoners crouching down. They squatted there, scanning the terrain as intently as a manuscript - searching the dirt for more stones.

Days later, surrounded by the unfurled rugs of a Kabul bazaar, a man examined the photograph I had given him. Chanting the rhythmic Pashto verses, he passed his hands over the creased and battered image of the Taliban love poem inscribed across that foundering wall.

He began again, in English:

You who judge me
I hope you burn alive and become dust
I hope you are destroyed and disappear from
this universe
Your days and nights filled with sorrow and
pain
Tear open my chest and see what is inside
Only then can you understand

Katrina

Hurricane Katrina had wound the entire Gulf of Mexico into her spiraling body. Reports of thousands drowned. New Orleans was an Atlantis.

I drove the twenty hours south, through all those places named after dead Indians: *Wyandotte, Erie, Maumee, Fort Shawnee, Wapakoneta, Tullahoma, Chattanooga, Chattahoochee, Chickamauga, Tuscaloosa, Eutaw, Ponchatoula, Shenandoah...*

A stolen Gideon Bible was my passenger. Police at the roadblocks saw it on the seat beside me. I lied to them. I explained I was looking for my sister, that nobody had heard from her since the levies broke. I was surprised that for the first time since I was a kid, I really began to cry.

I don't know where it came from. The police let me pass.

Jefferson Parish was a city abandoned. Just the alarms of looted stores and helicopters floating overhead like tadpoles in an upturned lake. A few hot rods raced down empty roads under dead traffic signals.

A bare-chested man with a moustache was smoking the tires of his Trans Am, burning black circles into the pavement of a parking lot.

Taking pictures? he yelled over.

Yes, I said. *What are you doing here?*

I work at that supermarket. He pointed. *It's a mess in there. They took everything.* He smiled like Christmas morning.

Thousands of mostly black refugees lined the highway, faces shining in the sun. There were enormous women too big to fit in any car, sitting with slippers and colorful plastic curlers in their hair. Some organization had dropped off a few fold-out cots and towers of water bottles as tall as a man, shrink-wrapped in plastic. In the shade of an overpass lay the elderly, the sick, and the dead, all mingled together.

Everyone had been waiting days for the evacuation buses whose drivers had all been

chased away by sniper fire.
At night, with the power out, for the first time in anybody's memory, they could see the stars. A little kid was hunting for the switch that turned them on. Families lying together on their backs - pointing, whispering, counting. Inventories were taken. Meteors were spotted and wishes were made.

The authorities had encircled the wealthy city of Gulfport in double cyclones of razor-wire. Even before the waters receded, the spray paint went up:

U LOOT I SHOOT

LOOTER HUNTING CLUB

THIS STORE IS BEING WATCHED

I SEE YOU .45 AUTO

LOOTERS SHOT ON SCENE

And a sign in front of one home, on a pile of destroyed possessions:

TAKE WHAT YOU NEED

At the ocean's edge in Mississippi I asked a man in a parked car, *Where are we?*

He answered, *This is Waveland. I mean, this was Waveland.*

Three men with metal detectors rhythmically swept the beaches. This had been a city of mansions. Just the tracings of floor-plans now. The only upright things: a bank vault with its slab door yawning and a chair on which 12 unbroken glasses stood in perfect rows.

Soldiers lounged under the few remaining inland trees. The leaves had been stripped away and replaced by clothing and plastic bags that cast colors down like stained glass. The soldiers and the Red Cross workers took turns smiling for pictures in front of a twisted railroad track, pretending to drive crushed cars, or captain beached ships.

A man poked through the sunroof of a passing truck, videotaping. A father came by ATV with his two tow-headed children on a tour. They smiled and waved at me. It seemed especially heartfelt. With the old unconscious habits of life disrupted, every action was an invention.

That night, I presented myself at a Pentecostal church in northern Mississippi. They had laid out a hundred beds for refugees, but nobody came. A flock of women fed me until I could eat no more. In return, I cried for them about my sister and they cried. I slept in the child daycare room behind a one-way mirror where parents could hide and watch their children play.

Makhmour

The village had no name. Everyone who had known the name of the village was now dead or had fled. When the Peshmerga had recaptured the settlement from ISIS that spring, it was so full of booby-traps that they just torched the place rather than deal with it. The town was abandoned now - just somewhere for the men to come scavenge.

This one's my house, Christopher Smith grinned. The former Marine corporal gestured with his battered AK-47 toward a fire-gutted jungle-green villa. All the buildings were like that – vibrant non-sequiturs of blue, yellow, purple. *It's like Super Mario World*, the 25-year-old remarked.

The village we walked through was slowly turning back into desert - disappearing by the truckload as its wreckage went to fortify the Mullah Abdullah frontlines two kilometers away. The end of Kurdistan is marked by a dull earthen rampart studded with the bright dreamland fragments of the nameless village. 700 meters beyond, across a minefield, is the Islamic State - or Daesh, in soldiers' slang.

Two months before that day in December, Smith had been a brick mason living in Vermont with a fiancée. That was all over now. *I took the wrong bus to Miami,* he joked.

After he landed in Sulaymaniyah, Smith called the FBI, notifying the agency of his whereabouts. *They knew I was here,* he insisted, *but they still felt the need to wake up my mother in the middle of the night and give her a heart attack.*

The American spoke of ISIS atrocities, the stories that had troubled his sleep back home: the beheadings and crucifixions, the slave markets, the rape camps - an evil that overwhelmed his senses. It was the reason he had come to fight, and like many of the volunteers I met, he was not here just to kill Daesh - he was here to send Daesh to hell. The foreigners hunted wild boar to supplement their rice rations and made sure that every round fired in anger was coated in the animal's unholy blood.

Hell is real here. The black inferno rages just beneath our feet, my translator warned. Everything is known. The flames have been measured and they are 69 times more painful than terrestrial flames.

There were always two wars going on - the one we could see, and the one we could not. Over 1,300 years ago, the prophet Muhammad spoke of the world's final hour. The armies of Rome, he said, will be lured to the plains of Syria and annihilated. Only then will the messianic Mahdi descend from heaven to defeat the cycloptic antichrist, Dajjal. Every phenomenon here is infused with mystical significance. ISIS distributes photographs of one-eyed babies, and the Pesh commander at Makhmour reports recurring nightmares of Kurdistan's capital underwater - a city of corpses carried deep into the earth by an inescapable current.

There were always remnants of that other ethereal war. The Shia mosque here had once been decorated with murals of humans and animals, until the jihadists came to paint blue circles over every face. Depictions of living things are considered by Sunni to be idolatry. On this wall, men with blue orbs for heads ride off into battle and a blue-headed raptor perches victoriously above his dead blue-headed prey.

Angels do not enter a place where there are images, my translator informed us. Then he started rapping, *I got my angels on my shoulders and a quarter of that angel dust.* The 30-year-old alternated continually between Quranic verse and Lil Wayne lyrics. *When I listen to Lil Wayne, I feel my faith decrease*, the pious Muslim lamented. *Sometimes I forsake him, but I always go back.*

Azad, a volunteer identified only by his nom de guerre, had departed the land of Lil Wayne eight months prior. He left his family, his job, and his given name back in Texas. The 46-year-old said a prayer one night and the Lord directed him to join the YPG - a group of communist guerrillas in Rojava.

You show up in a foreign land at four in the morning with a $150,000 bounty on your head, Azad recalled. *You're supposed to meet someone in an airport parking lot, but you don't know who. You're in the back of a car, and every time it stops, you're afraid you're going to end up on a milk carton. When you get to the safe house, you're feeling a little bit better, but only when those guys go to sleep - only when you hear them snoring, can you kind of relax a little bit.* In his sleep-deprived mind, Azad conjured up a series of frightening scenarios. Maybe terrorists had killed his intended hosts and taken their place. Maybe his hosts were planning to sell him to ISIS themselves.

After being smuggled into Syria, the Texan was sent directly to the front. *I was never combat trained,* he confided, *so I worried that I wouldn't be good enough. When I got there though, I realized that just by being an avid hunter I knew more than most of them did. My first thought was, Great, I'm not the weakest link. Then I thought, Fuck, I'm not the weakest link.*

Azad recalled fighting with the YPG alongside former members of the French Foreign Legion and the American military. There was also a man who claimed to be Special Forces but was actually a child molester. There was a Canadian lingerie model, and a Muslim convert who liked to kiss dead Daesh on the lips.

It's turned into a fan club, Azad grumbled, explaining why he had crossed back into Iraq and joined the Pesh. *A lot of people come over here thinking they're going to be John Wayne and Rambo running and gunning. It's not like that.*

Nicholas Barrett, a former combat medic and self-described *proud American infidel*, detailed a typical day in Kurdistan. *You'll have these commanders from different tribes who hate each other, and they're just klicks apart across no-man's land. They're both on the same radios that they bought at the same bazaar in Dohuk. They have their morning prayers, drink chai, talk about each other's mothers, and then lob mortars for a few hours.*

The men have to find their own entertainment. The morning I arrived at Mullah Abdullah, Christian Österman was cursing. A teenaged Pesh soldier had stolen one of his boots as a prank. With virtually no translators available, communication between the volunteers and their hosts was mostly hand signals and a smattering of Kurdish Sorani. *I told this guy that asshole is a sign of respect - like sir,* Smith said, waving at the boot thief. *Hi, asshole!*

The men kept busy building redoubts, scavenging firewood, spray-painting graffiti of Daesh soldiers copulating with goats, and caring for what remained of a litter of puppies birthed by a local stray. A rival bitch had been sneaking onto the base to the kill the puppies one by one, and now there was only a single scrawny mutt left.

At night, you could tell exactly where ISIS was by the glow of their cooking fires in the distant town. The Westerners would wait until all the little lights were glowing before flashing the positions with a laser pointer. They would laugh hysterically as the jihadists, fearing snipers, frantically doused the flames.

In the daylight, the town looked dead. There had been a black flag flying from a rooftop as recently as a week before, but it had vanished in a storm of French airstrikes. Today, the only sign of life came in the form of shrieking mortars, sent over at irregular intervals. *You can tell generally where they're going to land just by the sound,* Österman explained. The Swede had been in country for nearly five months. *When you hear that deep tone you know it's going to be pretty close.*

The men were so attuned to the signature trill that they sensed it through everything - conversation, engine noise, gunfire. By the time I heard it that morning, they were already on the ground waiting for the explosion. There was only silence. We raised our heads and looked around. It was a dud. The mortars are often bad and the ground is supple.

That evening, Smith and I hauled ourselves 100 feet up a radio tower to observe the enemy positions. The structure we climbed had been thoroughly scrapped of its most valuable metals by the Pesh who had not been paid in four or five months. The men used to have a telescope, but the volunteer who brought it had taken it home with him to America. Now, we just peered through the jumpy optics of Smith's AK at the floodplain stretching out before us, a still-life landscape of overgrown farms and ruined villages.

If I was in charge, the American reflected, *I'd bring in an EOD team to clear a path and send everyone up in one straight line as fast as we*

104

could go. We'd take that spot beyond the river, and from there we'd be golden. I'd set up a .50 cal right in this tower, because from here we could hit everything. Unfortunately, we only have about 100 rounds in the truck. We just don't have the resources.

As we prepared to climb down, Smith stood, chambered a round, and fired his AK into the vastness. The Westerners all had to buy their own ammo from the local arms market, and that shot had cost the American one dollar - about the same as a pack of Kurdish cigarettes.

When we hit the ground, the Frenchman Angelo Sammantano strode over. *That bitch ate the last puppy,* he reported, shaking his head.

Of all the men in that place, the American had loved those puppies the most. He had even made arrangements for an NGO to vaccinate one of them so he could bring the animal back to the States.

Sorry man, Sammantano added.

Smith patted his AK and glared at the contented killer lounging by the road. *We're going on a walk tonight,* he swore.

On the Marine's forearm was a tattoo of Atlas, straining to bear the impossible weight of a world engulfed in flames.

Is that you? I asked, pointing to the Titan.

Maybe, he said.

Meat

There is always one pig who's curious - always, Agnete Poulsen began. *And, when the gate opens, he'll be standing there looking around.*

She mimicked a pig's swiveling head as he steps from the dim hold of the transport truck to survey the florescent labyrinth before him.

Poulsen's voice playfully jumped an octave as she spoke for the pig.

What's going on? What is this place? Why are we here? Come on guys, let's go and have a look!

Then off they go, she concluded returning to her usual genial tone. *They go, and none of them register that there is something else happening just on the other side of the wall. They walk along as fast as they can - completely safe, and calm, and relaxed.*

As the head guide of the Danish Crown corporation, Poulsen has facilitated nearly a quarter-million visits through Europe's second-largest slaughterhouse in the ten years since it opened here in Horsens, Denmark.

We show 100 percent of what happens here, she boasted, *not 99 percent. It's the last 1 percent that makes us honest. We are living by killing pigs and you cannot romanticize that.*

Poulsen spoke of her upbringing on a local farm as we finished adjusting the elastic bands of our billowing hairnets.

Now we are alike, she quipped.

The tour began in the visitors' gallery. It was designed to move guests through the 21 acre facility with the same efficiency as the livestock below. Poulsen gestured at the stockyards beyond the glass.

Pigs have a group mentality, she remarked, *so, we keep them with their friends from the farms. They prefer to go upwards instead of down, so the floor has a two degree incline. They don't like to go straight, so we have curves along the way. You'll also notice that our workers wear green and blue clothes.*

There is no white, because pigs are prey animals and white reminds them of their predator's teeth.

Every day, 20,000 pigs enter the slaughterhouse under their own power, only to exit hours later as shrink-wrapped cuts of meat. The facility was devised to encourage the animals to move willingly. Electric prods are banned in Denmark, so the only driving tool available is a paddle, rigged to make a slapping noise when swung.

Poulsen was quick to point out the culinary benefits of Danish Crown's low-impact methods.

Stressed meat is actually bad quality, she explained. *It has bad texture, bad color, and it gets juicy in the wrong way. It's like chewing into cotton.*

Most of the pigs we saw in the holding pens were sleeping. Poulsen called the cuddling and gentle-biting we observed *their way of getting close.*

Is there meaning to the sounds they make? I asked, referring to the sporadic waves of squealing.

No, there is not, she replied.

We moved on to the killing zone.

This is where it all happens, Poulsen intoned. *The beginning of it all. The beginning of the end.*

The glass made everything seem farther away than it was. We were above and off to one side - the perspective of an out-of-body experience or a third-person dream. I had never noticed before how delicately pigs walk on such small hooves. All that bulk and they don't even sway. I had a flash of elegant women in high-heels. I saw parties, and dances, and little courtship glances - life in all of its disguises. Here was life wearing the mask of a pig.

In my memory, the pigs seem to materialize out of a fog. There was a man in blue or green leaning over them - close enough to pat their heads if he had wanted to. It was almost as though he was just there to say farewell, but it must be that he pressed the button that sections off the next group of eight. I don't remember his face. He might have been wearing a mask. A gate rises. The pigs walk under their own power until the very end - those just-so steps. The black ribbon moves forward, and they are pushed, gently but inexorably into the dark elevator.

Poulsen described what I could no longer see - Danish Crown's method for stunning their animals before slaughter.

We send them down nine meters below the ground - down into the chamber with the carbon dioxide. They fall asleep, and after three minutes they come up.

I spoke later with Dr. Temple Grandin, a leading meat industry consultant and designer of livestock handling procedures.

What do pigs experience during carbon dioxide stunning? I asked.

I've seen a lot of variation, Grandin replied. *What reaction you get depends upon what genetics of pig you put in there. The ones with the Porcine Stress Gene have a bad reaction - they panic.*

Denmark, though, she added, was one of the countries that had bred that detrimental gene out of their livestock. She described a CO_2 induction that she had witnessed in the country.

The pigs had some backing up and sniffing, and then they rolled over and started convulsing. Once a pig falls over and goes into convulsions, the pig is unconscious.

Back at Danish Crown, the gassed pigs emerged out of a chute in the various abandoned postures of slumber. A man deftly attached shackles to the pig's hind legs and the bodies rolled up into the air. They swayed then, gently, as they glided to the slaughterer. The man pulled a sapphire-colored tube down from above. At the end of the tube was a blade.

The knife goes through the trachea into the big veins around the heart, Poulsen narrated. *Then the heart itself pumps out the blood, and then it dies. The blood flows up the tube and is separated into protein and plasma. You can use it for human consumption, animal consumption, or medical use.*

Every part of the pig is used, Poulsen boasted.

The hair is for brushes, the skin is for gelatin, the manure is sent to a biogas plant to make energy, and the warmth of their bodies is used to heat water.

On our way down onto the slaughter-line, we shuffled through a gauntlet of mechanized cleaning apparatus.

One of the largest hospitals in Denmark visited to see how we control bacteria, Poulsen bragged. *We have cleaner hands than a surgeon.*

In a mist of blood, a massive circular saw split each carcass down its spine. Inverted pig-halves floated in with a lazy, clapping sway as the slaughter-line workers churned around them in muscular orbits without a wasted gesture. The machines set the pace for a dance

with planted feet. A cacophony of discordant rhythms synchronized in flesh as animal organs spilled out in jewel-glazed bundles. Their insides look just like ours.
You can replace most of the human body with parts from the pig, Poulsen remarked. *In another of our slaughterhouses they take out the heart valves of sows and export them to America for transplant.*

Tiny radio chips embedded in the ears of each pig coordinate with ultrasound machines to automatically scan and sort the bisected pigs according to meat quality and fat distribution. Computers plot out the most effective cuts for the automated saws to perform, as each carcass is gracefully routed by overhead conveyor to its correct location in the plant. This precision, along with a flexible workforce, allows Danish Crown to fulfill the demands of customers around the world.

For Japanese products, Poulsen noted, *we use four people to brush off and Hoover the meat. For Danish products, we use none. In Japan, the customer is so far removed from the product that they find something like a bone or a small piece of fat on the belly to be a foreign object. Of course, it costs extra, but we will always adjust to the customer demand. We replaced an electric guillotine robot with three workers just to be able to deliver heads to the Chinese market. The Chinese market wants split heads.*

As the carcasses are processed, they look less and less like the animals they once were. Feet, tails, and ears are removed in stages. Once the faces with their infantile eyelashes are gone, all that is left are cuts of meat.

That's how it was by the time we got to the American line. The American line is where babyback ribs come from, Head of Media Relations, Jens Hensen told me over the phone several days prior. He sang a few lines from a popular restaurant's television jingle, *I want my babyback babyback babyback...* I could hear him grinning through the receiver.

The burly butchers on the line were grinning as well when I pulled out my camera to film the plant's fastest worker.

He's a pretty little thing, Poulsen joked. A couple of the butcher-women agreed.

That's the most beautiful guy of the lot! one shouted.

You can put me on the table next! the other called out to him.

He was blushing like a teenager as he slammed thick rib slabs onto a counter before trimming and deboning them with the precision of a surgeon and the speed of an athlete. It was piece-work, so the man was racing, but his body was so accustomed to the demands of the process that he wasn't even sweating.

We talked to another of the butchers, a seven-year veteran named Daniel. His arm was encased in a Medieval-style chainmail armguard that rattled as we shook hands. I asked about the tattoos covering his bulging biceps. One was a portrait of his dog with banner beneath proclaiming her *Queen of the Streets*. He talked about how she loves the raw meat smell that clings to his body.

She always licks my hands when I come home.

Another tattoo read, *You can't change the past, and if you want to predict the future, you have to create it.*

Poulsen cited the tattoo, *Did you notice how it references the past? There are a lot of people here who have had some difficulties in their past. We've actually had prizes for giving people an extra chance.*

We said our goodbyes and watched Daniel return to the line. The personal rhythm of his walk gradually recalibrated itself to the movements of the men beside him - the universal rhythm of the plant. It was my rhythm too, as I darted through the periodic gaps in the endless moving line of hanging carcasses. Poulsen dreamed of the rhythm.

When you go to sleep you just keep seeing the meat moving.

Book IV

Tongji!

As once it appeared over Bethlehem, so too in North Korea - a new star scorched the heavens. A storm broke. Twin rainbows sprung forth. In sonorous human voices, soaring swallows spoke the name. On the face of Mount Paektu, in wreaths of flame, blazed the name.

Glorious Leader Who Descended from Heaven!

Highest Incarnation of the Revolutionary Comradely Love!

On February 16th, 1942 - The General Kim Jong Il was born.

Gunfire and the sounds of revolutionary struggle were his lullaby, added the beautiful Ms. Song. *And many such natural miracles also happened when The General passed away last year. In the moon, for example, we saw his picture.*

Ms. Song was the official minder assigned to me throughout my stay in North Korea. The face of her country - in style and manner she resembled a miniature Jacqueline Kennedy.

They flew me into Pyongyang on an Anatov An-24, a Soviet turboprop last produced in 1979. Beneath its wings: vast fields, low mountains, houses in tight clusters. No dwelling stands alone. The center of every village is marked by the stone pillar that reads:

THE GREAT LEADERS KIM IL SUNG AND KIM JONG IL ARE FOREVER WITH US!

Magazines were distributed by the flight attendants. Page 1 featured an article on the activities of the new 28-year-old leader Kim Jong Un as he toured an amusement park:

He familiarized himself with the operation of Z-Force, Power Surge, Pirate and Volare and called at the electronic amusement hall, kindly asking which apparatus the people preferred and underlining the need to set up the chair plane...

Page 16 comprised a series of statements by members of the Korean Children's Union. A young girl, Ri Kuk Hwa:

I felt like in the dream when I, daughter of an ordinary postwoman, presented a bouquet of flowers to Kim Jong Un. When he took my hands in his, I found in him the image of the sun. I will keep in mind his benevolent image and remain faithful to him like a sunflower following the sun.

Another young girl, Ko Hu Hoe:

About 10 years ago my family had the honour of having a photograph taken with Generalissimo Kim Jong Il, who came to the high and rough pass in Jagang Province. At that time I, as a baby, slept in my mum's arms. Thus my grandma named me Hu Hoe (regret) and I have always regretted for having slept that day.

At the end of the flight, the attendants returned to collect the magazines. Ms. Song explained that no paper featuring pictures of the leaders is ever thrown away. Instead, they are collected and respectfully stored in vast warehouses forever.

So revered are the images and words of the leaders that a girl who drowned during the recent floods trying to save two of their portraits became a national hero. A museum was erected to display the singed clothing of 17 soldier-martyrs who burned to death in a forest fire protecting a stand of slogan-trees on which were written words praising Kim Il Sung. It was said that the soldiers' blackened arms had to be pried from the trunks.

It was soldiers like these who met me as I disembarked. I glanced down at their drum-tight faces, swaying decorations, halo-high puffed-out visor caps. Informed that I was an American, the senior officer addressed me via translator:

Korea and the American Imperialist Aggressors are still at war. The agreement signed in 1953 was only a cease-fire. The Americans may feel safe, hiding behind a vast ocean, but vaster still is the determination of the Korean people, to whom an ocean is only a puddle. Remember this always.

At immigration, they feasted on my luggage - waving my shirts like flags, peeking at each other through my glasses. Carefully, they studied my notebook, page by page, line by line, upside-down. And finally, as if it were a trophy he had won, a smirking soldier hoisted my contraband: a ceramic statue I had bought in Beijing. It was Chairman Mao.

The soldier scowled and shook his head.

Mao, I said. *That's Mao.*

The soldier obviously believed it was some kind of religious icon. He drew back his lips, luxuriously exposing his teeth.

Mao, I said.

The soldiers crowded in. I felt a hand grasp my collar and my shirt tighten. They were shouting for someone. I waited, breathing through my collar, until the young soldiers parted and an old soldier joined me in the center.

Ah, he said, appraising the statue. *Mao Zedong.*

The old soldier waved his hand and the pressure around my neck abated. The soldier with the statue, already bored, pushed the idol at me, and quickly turned away.

They hustled me into a van bound for the city center. The curtains were left open, but photographs were not permitted. Photograph no countryside, no fallow field, no tree stump, no barbed-wire, no idle worker, no soldier, no soiled clothing, no empty road, no peeling paint, no checkpoint.

Why are there so many checkpoints? I asked Ms. Song.

For national security - against spy, she replied. *Didn't you see the interview of the South Korean spy? They caught her trying to explode a monument. It was on the television station last week.*
Oh, we don't get that station back home.

Really? I thought all countries get that station.

As we waited at another checkpoint, I pulled out a small phrasebook I had been given. *Let Us Learn Korean* the title cheerfully suggested. In this way, I learned to greet everyone with a hardy, *Tongji!* That means, *Hello, comrade!* And then I might ask, *Sugo hasimnida?* which means, *How do you do?*

If somebody asked me, *Sugo hasimnida?* the phrasebook gave two choices of response. I could say, *Josumnida,* meaning *Good,* or I could say, *Cham josumnida!* meaning, *Very good!*

For any contrary sentiment, one was left only with the medical section:

I feel nausea.

Or perhaps:

I have a toothache.

We sped through the city center. No traffic to slow us, but traffic-girls at every intersection in smart blue uniforms - girls hand-picked for their beauty, waving us through. Crowds on the march. Children drilling for torchlight parades. And everywhere, pastel-painted apartment blocks, balconies overflowing with flowers. Now you can take pictures.

Other useful phrases:

Pyongyang is clean and beautiful and seems to have the best housing conditions in the world.

All the Korean people are good-mannered, diligent and modest.

Korea is the people's paradise where there are no beggars and all people study.

There were no advertisements, no seductive women pushing products, not even the outline of a breast - only the hand-painted blood-red signs with the white lettering:

WE WILL DO AS THE PARTY INSTRUCTS!

WE HAVE NOTHING TO ENVY IN THE WORLD!

LET US CARRY OUT THOROUGHLY THE BEHEST OF THE GREAT LEADER KIM IL SUNG!

Every now and again, in black lettering, like a cavity - the word *American* appeared.

WE WILL DESTROY THE AMERICAN IMPERIALIST AGGRESSORS!

We passed a mural of American soldiers throwing bound Korean women into a river.

There were paintings of dead American soldiers clutching dollars. And, in the amusement parks, over and over, painted onto boards - that man with the beak-nose, the beady bloodshot eyes, the fangs, his limbs blown off, engulfed in flames: The evil American G.I. Shoot a BB through his head and win a prize.

There were these things and then there were the smiling, paternal faces of the leaders. Thousands of Kim Il Sungs and Kim Jong Ils gazing lovingly at us from all directions. Their portraits - in every building and pinned to every chest. They stood larger than life on every block in murals and as statues (never hollow, always solid metal). They pointed into the distance at something we could not see. They galloped stallions over mountain peaks. They guided farmers at their plows. Children streamed into their expansive embraces. They smiled the most compassionate, white-toothed, eye-crinkling smiles.

And the animals smiled. The concrete owls, deer, cartoonish bears, frogs in suspenders, accordion-playing rabbits in skirts - laughing and grinning in defiance of anatomy. The menagerie - in parks, in front of buildings, lining the roads. The animals stood among fake miniature mountains and giant toadstools that doubled as parasols for crouching soldiers.

We drove across the nation in our official tourist van. We stepped where the leaders had stepped, footprints carefully inlaid with tile, to waterfalls with heights that echoed their birthdates, to gardens brimming with their namesake orchids and begonias, to nurseries and orphanages where the children sang their praises and danced hoisting models of their nuclear missiles. We laid flowers and bowed before their monuments, their birthplaces, their tombs. We drove upon their roads.

We passed few vehicles traveling down the Young Heroes Motorway. All 42 miles, all 10 lanes of it were built at Kim Jong Il's behest, Ms. Song announced proudly. It was constructed by youths, by hand, and all of it during the height of the famine just to show the unshakable single-hearted unity of the Korean people under adversity.

The famine is finished now, she insisted, as we arrived at the nation's largest apple orchard. It had been commissioned by The General. He had visited three times, to stand where we stood and look down on the acres of bountiful trees enclosed by high fences, with armed guards every hundred meters, scanning the empty countryside. The trees, in their neat rows, had reminded The General of soldiers on review.

We visited the beach at Nampho. Under playful protest, Ms. Song had donned a frilly one-piece bathing suit and red rubber cap to join me in the surf. We threw a Frisbee back and forth. When it landed short and splashed her, she smiled as she swore, *American imperialist!*

Whenever I teased her, she would laugh and threaten to throw me in jail. On two occasions, she bumped her head as she ducked into the van. I offered to buy her a helmet.

What kind of helmet do you want? I asked.

U.S. Army helmet, she grinned.

You already have plenty of those in the basement of your war museum, I chided, *down with all the rest of your trophies - the burned out tanks and planes.*

Those helmets are all worn out, she replied. *I want a new one.*

We climbed all afternoon and stood on the peak of Mount Paektu.

I don't care you're American, she said finally, breaking the silence.

I didn't know what to say. I said, *Thank you.*

As we scrambled down like goats across a narrow span back to the trail, she glanced at the valley floor hundreds of meters below. She reached out and grasped my hand. Just the last three fingers at first. I didn't see her do it, only felt the rough, calloused skin of her palm, and the firm pressure of her little fingers. I pulled her across and we stepped down onto the trail. I waited for her hand to slip away, but she didn't let go. I couldn't understand why she kept holding on, but her hand fit just right and we walked that way a while, winding down the trail before our fingers jangled loose.

In the west, I said, *it's believed that if someone speaks against the government here, they will be badly punished.*

In the west, Ms. Song replied, *the people are undisciplined in their mind and heart. Here we have one strength that is not to be found in all the world - that's the single-hearted unity. The single-hearted unity means one thinking, one ideology. Every people think same. Every people do same.*

But if they didn't think the same, I insisted, *they would be punished, wouldn't they?*

She hesitated for a moment.

But, it cannot happen in this country, I think. One cannot have a different idea, because the government that is represented by leader Kim Jong Un, Kim Jong Il, and Kim Il Sung, gives the policy that is very very right and that is very welcomed by the thinking people. It is very natural to the thinking people's idea.

Everybody likes it. For many westerners it's difficult to understand this reality.

The apotheosis of this single-hearted unity is perhaps best demonstrated by the Arirang Mass Games. Every autumn, in the world's largest stadium, 100,000 extravagantly-costumed participants perform highly-choreographed gymnastic and dance sequences glorifying the regime. 30,000 school-children form the backdrop, each holding one of a series of colored boards, deftly swapped to display enormous pictures behind the sea of performers.

The sea - the gymnasts and the dancers: individuals that congeal into lines, become shapes, become patterns - each one highly-skilled, and yet, each one a perfect replacement. There are no soloists.

As we merged into the river of people streaming out of the Rungrado May Day Stadium, Ms. Song turned to me. *When I was a girl, I danced Arirang,* she whispered dreamily.

What part did you dance? I asked.

I was a leaf, she said, extending her arms, bending slightly at the waist, this way and that.

Where? I asked. *Where were you?*

Beneath the flower, she replied. *Could you see? Beneath the flower petal is the leaf. Beautiful green dress of the leaf.*

9/11

A letter from Aldo Brow:

As I write this, I am in a tent on my bed to avoid the cameras that are all over my apartment. I have been chased around the world. I have survived over 1,000 attempts on my life.

It is Saturday morning, April 19, 2012. I currently have an NSA husband and wife crew living above me. All of the apartments on my floor have been filled with NSA agents. Mel in 24-C and his friend in 24-D have been recruited by the NSA. Also, Elaina the real estate broker, who owned the pug type dog, is NSA. All employees of the Galaxy Apartment building and its management team have been recruited by the NSA.

My microwave cannot be used because it is giving off microwaves outside of the oven. My cell phone is also giving off microwaves same as the oven. My computer and TV are also giving off these waves, along with my electric shaver. This frequency was used in the 1991 Gulf War to terrorize the Iraqi troops' minds and cause over 100,000 to surrender.

I have a GPS chip in my tooth, credit card, and driver's license. Whenever I go out I am tracked and followed by the NSA.

I know you are thinking, why have I survived? Why haven't they killed me yet? The reason is, I discovered the secret identity codes and use them nonstop to avoid being killed. The NSA has a secret identity sign to alert other agents that you are one of them. This is how I survive...

Aldo was wallpapering Times Square when I met him - staple-gunning big xerox broadsides. He was clutching one that read:

ALDO BROW WITNESSED THE NSA ARRANGE 9/11

He saw me watching. *Do you work for the NSA?* he asked.

I told him I didn't.

It's okay if you do, he continued. *The NSA is a*

great organization. The people in the NSA are great people. But there's a few people who control the computer in the NSA, who are moles, actually. Kruschev said he was going to destroy America without firing a shot. I'm not saying it's Russia. It could be the Chinese. It could be someone who's gone mad in America. I don't know...

Aldo was stocky, but he was elfin. A man in his forties with an uncreased, untroubled face. He had the big blue eyes of a calf. Big open eyes, talking like a sugared-up kid, words leap-frogging.

Why would the NSA arrange 9/11? I asked.

When you create fear in the people, he explained, *you can get them to give up their rights. The Patriot Act was presented to Congress and the Senate... It took away our civil liberties, our privacy. Now they don't need a warrant to listen to every phone call, so now they listen to Wall Street and buy and sell in every stock, and now they have unlimited budget, unlimited power. They can monitor anyone, kill anyone.*

Anyone that speaks up better not go to any restaurants. The NSA calls the chef who puts poison oil in your meal - looks like a heart attack. The oil is toxic but it tastes fantastic.

It was Aldo's sister-in-law, Livia, who had orchestrated the September 11th attacks. Six months prior, Livia had gotten drunk at her own wedding and blabbed about plans to crash planes into buildings. She mentioned the name Khalid Shaikh Mohammed. She claimed to be NSA.

I didn't believe her at the time, Aldo noted, *but the NSA must have been watching because after that, everything changed. Cars started blowing up, child services would show up, people tried to break into our house to kill us every night. I mean, every night someone was trying to come through our windows or our doors. At first I thought it was the repo-man. I had everything barricaded, booby-trapped.*

My wife Vivian could not understand. She got nervous. Things were just going haywire. She took the kids and went back to her mother's. It was a good thing they left cause they could've gotten killed.

That was the last time I saw my wife. I heard her voice on the phone, but it wasn't her. The NSA can imitate anyone's voice. I believe that they either replaced her or were using some kind of brainwashing on her.

I have affidavits that my wife's blood type and foot sizes have changed. The experts tell me that they grab the original people, lock them up in a facility like Area 51, and force them to transmit answers to a body-double wearing an earpiece. The body-doubles - they put a mask on 'em, like a Halloween mask, or do plastic surgery to make 'em look the same. Nano-technology voice-box makes their voice sound the same.

So after all of that, I asked, *what did you do?*

The only thing I could do, Aldo said. *I decided to run for president. I am determined to let the world know what really happened on 9/11, and to get my family back if they are still alive. If you spread the word about me, I'll make you my press secretary.*

I'm pretty sure you work for the agency, he continued, *but if you don't work for the agency, they're going to approach you in a little while and give you money to keep quiet, okay? If you take the money, enjoy it. Find a girl, buy her a nice ring, take her on a cruise, take her to Paris. They've got EPCOT there, they've got Disney World there, good dining, beautiful surroundings, and it's nice - if the agency will give you time off work.*

Thanks, I said, *but I really don't work for the agency.*

Well, he insisted, *you will soon.*

Chaplin

Every morning when I get up, Dr. Ashok Aswani began, *I just remember Charlie's face. I pray to him as my God. Don't let me cry. Because, if I laugh people will laugh with me. Make me like you, so I can make other people happy.*

The 68-year-old gestured towards a shrine. A ceramic statuette of the Little Tramp stood surrounded by a pantheon of figures: Jesus, Buddha, Gandhi, Hanuman, Ganesh.

We arrived as strangers in the city of Adipur, India. We asked for Aswani, but no one knew the name. Then, we asked for Chaplin, and everybody pointed. Last month, in search of the doctor, a woman in a taxi had simply flashed a crude sketch of a mustached man in a hat.

As Aswani's wife later said, *Ashok is Charlie.*

The doctor sat in his dispensary. In his role as the Ayurvedic doctor, Aswani was handing out packets of medicine and prescriptions for *The Kid, Gold Rush*, and *City Lights*.

You look like Charlie! I laughed, when I saw him. He had smirked at me in that familiar way – eyebrows darted, shoulders flaunted, lashes fluttering.

You look like Charlie! he echoed.

He turned to my translator, *And you look like Charlie! Your height, those eyes, that smile.*

There is some Charlie in everyone, he added.

Nowhere is that more apparent than in Adipur. Every April 16th in honor of Chaplin's birthday, the entire town parades around in black hats, stub moustaches, and bamboo canes - men and women, young and old.

The celebrations were not always so extensive. They began in 1973 with only Aswani, his wife, and two sisters.

At first, Aswani beamed, *the people couldn't even pronounce Chaplin. They called him Charlie Champion! Now, during the festival, for two kilometers you see only Charlies.*

Charlies laughing. Charlies dancing. Charlies singing.

Among them, is professional Chaplin imitator Jason Allin, who flies in annually from Toronto, Canada. *They draped me in flowers,* he recounted. *They kissed my feet.*

The crowd capers past an empty plot of land, disturbing the weeds. The doctor has already built a temple to the film star here in his mind. When the stone rise, Aswani insists, the halls will be full of worshippers, crooning their plaintive Gujarati hymns:

Oh Charlie, we are crazy for your name
You always smile and make us smile
Black hat on the head
And black coat on the shoulders
Here you come shambling
The image of you, Charlie
We will welcome you so well
That you will forget your London

Chaplin is always referred to here in the present tense.

He is not actually dead, Aswani's daughter Monica Navani explained. *He didn't die. He just went away.*

For Navani, growing up, Chaplin was always present. *Charlie is part of the family*, Navani reminisced. *Dad always used to put on his movies. When you come home, you see your father, your grandfather, and you see Charlie. It's the same. Those movies are as good as an autobiography.*

They spoke the language of silent film. *We never actually used to talk a lot, my father and I*, she continued. *We used to communicate by making faces and smiling.*

In the religion of Charlie, smiling is prayer.

I have known Ashok a long time, fellow Chaplin imitator Haresh Thakkr observed. *He has had many difficulties in his life: social, physical, psychological. But, he never shows anybody. He is always happy. He is a friend of life.*

We drank tea at Aswani's home, as he struggled to shave. Arthritis has curled his hands. When his sleeve fell past his wrist, we could see the beginnings of the deep scars that crisscross his body. A fall, several years ago, between the cars of a moving train nearly killed him. He is able to walk now only with artificial joints. It has become impossible for the man to imitate so many of his idol's signature gestures and shuffling gait - movements that a young Aswani had once practiced for 6 to 7 hours a day.

Grandson Talin Navani stepped in to finish the shave before anointing his grandfather's upper lip with a little fake moustache. The older man immediately brightened. His body seemed lighter, as he stood, grabbed his hat, and led us to the roof.

It was there that the transformation became complete. With all of Adipur at his feet, Aswani balked, minced, simpered, and flirted with the air, just as his god had done nearly a century ago.

The people still don't know what Charlie was, Monica Navani lamented. *We have these beliefs that God does miracles. Watch Charlie's movies. Watch Modern Times. Watch The Circus. He is tightrope walking, skating, doing gymnastics, singing. How can one person do all those things? Isn't that something not from this world? He is like a perfect one, and the perfect one is only God. Even in being imperfect he is perfect."*

His state of perfect imperfection exists because of his vulnerability, she explained. *He is vulnerable to the kind of societal evils which we take so lightly, and which he shows in such a beautiful way. These things do affect us, but we carry on, leaving our innocence behind. He carries his innocence with him.*

He reminds me of Krishna, I reflected.

Yes, Navani confirmed. *Krishna would do pranks all the time, and he has innocence also.*

She told a story about Krishna as a child that almost could have been performed by the Little Tramp himself.

Even though Krishna was the son of a king, she began, *he had this habit of stealing butter from the ladies of the village. He used to wait until they went to work in the fields and break their earthen pots, and eat the butter, and waste it. One day, the ladies caught him and dragged him home. His mother did not believe their accusations, so they said, You can look in his mouth and see the butter. His mother told him to open his mouth. When he did, she could see the entire universe inside.*

When you talk about Chaplin's perfection then, I offered, *isn't it our perfection too that he's showing us?*

Yes, she replied. *Everyone is a vulnerable, innocent person on the inside, afraid to reveal what we are. He is simply showing us that.*

Earlier, Aswani had shared a recurring dream. He is in Switzerland, standing at the graves of Chaplin and his wife Oona. *I'm there,* the doctor marveled, *and I'm seeing them. They say, Ashok has come. They're just embracing me, and I'm crying on their shoulders. They say, You're doing a very good job.*

Śarīra

When a man of 80 kilos is cremated, he becomes 2.5 kilos of ashes, Rinaldo Willy explained, *with these ashes, we make a diamond of 0.2 grams, smaller than a button on your shirt. How heavy is the soul - if we have a soul?*

In its coupling of the tangible and intangible, it is a question that epitomizes Rinaldo's work. Every year, his company Algordanza, receives more than 800 urns filled with human ashes. For between $5,000 and $20,000, the contents of each parcel are transformed into a diamond.

It is also more than a diamond.

Maybe soul is too strong of a word, Rinaldo continued, still struggling to define the essence of his product. *Our process is purely physical -*

but if the deceased had blue eyes, and the diamond turns out blue, you can be sure that the family will say, Oh, it's exactly the color of his eyes.

We were sitting on the cool leather couches of Algordanza's simple reception room in the sleepy town of Chur, Switzerland. High in the Alps, the town seems isolated, and yet, the Indian Ocean tsunami, the Chilean earthquake, the Fukushima meltdown, and the terror bombings in Madrid have all sent ripples through Algordanza's halls. Within weeks of a major incident the parcels begin to arrive.

We had a British soldier from Afghanistan recently, Rinaldo mentioned. *He came home and then he came to us. His body - not him, of course.*

The route from the Chur train station to the company's facility passes through medieval cobblestone streets, a golf course, and wildflower fields. It is a journey that many grieving clients make.

We ask that the family either brings the ashes or picks up the diamond in person, explained Rinaldo. *For us, it's important that they see who the people taking care of their loved ones are.*

The pilgrimage to Chur is just one part of a choreography designed around the six-month gem-making process. As one of the first companies to enter the memorial diamond business a decade ago, Algordanza, whose name means *remembrance* in the local Romansh language, has developed a tradition all its own.

I told my staff that they are not allowed to make condolences at the beginning, Rinaldo said. *You don't know the people. You don't know their story. It's not honest. During this process, however, we inform the clients every time we do something - for example, when the chemical analysis is done, or when we start the growing process. So, if you start to form a certain relationship - if you make chitchat, and you start to learn who the deceased was, how he died, and who the relatives are - if you feel that you want to make a condolence then, you may, because it's honest.*

Other protocols include standing outside as the family departs until they are out of view, and delivering the finished diamond by hand inside of a polished wooden box like the one on the table before me. I watched as Rinaldo slowly donned white cotton gloves, and in a series of precise gestures, unfolded the box without a sound. It opened like a flower to reveal the diamond inside on a little pyramid.

It is special to me when I am able to deliver a diamond in person, he confided. *We do it in the living room or the kitchen with everyone around the table. It's a very emotional moment when you are returning a family member who was away for six months. The diamonds always bring back beautiful memories. If there are tears, they are tears of happiness.*

In the laboratory down the hall, the gloves came out again.

We never touch the ashes or the diamonds with our hands, Rinaldo explained. *It's too intimate for us.* He gestured towards a row of white canvass covers. *During the process when we're waiting for the next step, we always cover the remains so that they're not naked. We do this because we believe that's how we would like to be treated - not as a material.*

Each set of remains is assigned a reference number, both for discretion and for the emotional health of the employees.

It helps the people working with the ashes to have a certain distance, Rinaldo explained. *For me, the French are the most difficult. They have this philosophy to send a photograph of the deceased along with the urn. It's difficult to see a girl of nine years. What has she seen of this life?*

In accordance with Rinaldo's principles of dignity, Algordanza refrains from accepting pets, adding extra carbon if there is not enough to make a diamond (except in the case of infants), and artificially coloring their gems.

Technically, we could make diamonds that are yellow, green, blue, or red, like our competitors do, Rinaldo insisted, *but we believe in no manipulations. As soon as you have additives, there's something in the diamond that doesn't belong.*

Instead of being predetermined, the color of each Algordanza diamond results from the specific combination of trace elements present in an individual's body. Fake teeth, titanium hips, or the remnants of chemotherapy can all impact color. Nitrogen lends a yellow hue. Traces of phosphorescent chemicals can produce diamonds that glow in the dark. The blue cast that so often reminds families of the eyes of the deceased is the result of boron in the ashes, though an excess will turn a diamond black, as it did in one recent order.

I had an older gentleman calling me in tears, Rinaldo confessed. *He said, I don't understand. My wife was not a bad person. People always associate the color of the diamond with the characteristics of the person - black diamond, black soul. Go ahead, try to explain to a man in that situation that his wife is not a bad person.*

Despite the occasional disappointment, Rinaldo maintained, most clients are grateful for the service his company provides.

Memorial diamonds are changing the way we mourn, he explained. *When you bury someone, you always have a bad conscience. You have to visit him at the cemetery. Nobody likes to go to the cemetery. It's a negative association. You start to imagine him under the earth with worms in his body. If you cremate somebody, it's black and it's dirty. Diamonds though, have always had a positive association. We turn dirty ashes and bones into something beautiful. Instead of feeling loss, you can remember that person's life.*

Most of Algordanza's diamonds are eventually crafted into jewelry, worn by clients who want to be able to carry the presence of their loved ones with them at all times.

Many people talk to their diamonds, Rinaldo remarked. *If a wife wears her husband's diamond in a necklace, there are the usual jokes: He was always wishing to be between my breasts. He wanted to be close to my heart.*

Some customers bury their diamonds in meaningful places. One widower threw his gem into the lake where he likes to fish.

There was another older man, a farmer who was dying of cancer, Rinaldo recalled. *He said,*

When you make me into a diamond, just bury it in the backyard. One day when somebody finds me - can you imagine how happy that person will be? I said, Lutzi, you're crazy. But I thought it was beautiful to be confronted with death and still think about the happiness of others.

As we talked, the director led us out a back door. We strolled through the mountain fog towards a soft humming sound. The hum intensified as we entered the building that houses Algordanza's three diamond presses. Day and night they buzz with the quiet violence of the forces they replicate - transcendental cataclysms deep inside the earth. It is here, at temperatures reaching 2,500°F and pressures of nearly 800,000 pounds per square inch, that the carbon extracted from human ashes is transfigured into diamonds.

We paused to listen. There was a cadence to the droning that sounded almost like chanting. The 18-ton machines looked like enormous idols. Rinaldo laughed when I compared them to Aztec gods.

This could be a kind of temple, he considered. *We hope that in five years we can construct a new building here, maybe a cathedral for the machines.*

We talked about how jewels are used as spiritual metaphors in so many of the world's religions. I mentioned the Buddhist belief that pearl-like objects called *śarīra* can be found in the cremated ashes of spiritual masters.

Yes, Rinaldo smiled. *We had a party of South Koreans visiting who offered us some of those gems to analyze. It was clear to us that someone must have slipped them into the mouth of the corpse, or maybe the person swallowed them before he died. We can prove chemically where gems come from, and these were clearly from a mine. They were not organic. It was interesting to us though, because it helps us understand the way people think - if you are a good person, then in your ashes, you leave a gem.*

Patient 106

At the entrance of the Cryonics Institute, in wooden frames, they gazed at us: photographs in two neat rows. Many of the elderly had chosen pictures of themselves from the prime of life - the way they wanted to be remembered. The way they would someday be again.

With dreams of immortality, a math professor, Robert Ettinger, started the Cryonics Institute in 1976. In an unassuming building in Clinton Township, Michigan, the dead are frozen in liquid nitrogen until future generations develop the technology to bring them back to life.

Pointing to one of the portraits, caretaker Andy Zawaki explained, *That lady is Mr. Ettinger's mother. She was the first patient. The second patient was Mr. Ettinger's wife.*

Ettinger himself became patient 106 at the age of 92. *He got sick,* said Andy. *He just kinda wore out. They hired nurses to sit with him around the clock so we would know the moment he died and start the cooling.*

Along with the Ettingers were other families, and even their pets: dogs, cats, parrots, a hamster named Leapy. In another part of the facility were fire-proof file cabinets filled with the patients' possessions. Like pharaohs bundled into well-stocked tombs, everyone planned to resume their lives exactly where they had left off.

You know, said Andy, *people talk about the end of the world coming. Well, the end of the world comes every day for the people that die, cause once you die, nothing matters anymore. It's game over, man.*

There's a friend of mine who's frozen now. He had a stroke. He was only 77. He used to say that when you get buried you know you're gonna rot, and if you get cremated, you're gonna be ashes, but when you're signed up to be frozen, you don't know. There's just the little bit of a possibility of hope, and for my friend, that changed his whole outlook.

$28,000 buys a place inside one of the massive metal vats, called cryostats. Each cryostat holds six patients, upside-down, swathed in sleeping bags. Encased in white insulation, the cryostats look like mushroom stems. The storage room holds nothing else except a small row of cubbyholes, numbered for anonymity, where loved-ones can leave flowers.

The laboratory has a science-fair quality to it. Everything has been assembled by hand.

A lot of people come in and tell me they expected lights, and dials, and people monitoring, but we want it as simple as possible because we don't want to rely on all that stuff.

Andy lives at the Institute. He sleeps on a fold-out couch. Sometimes his girlfriend drives in from Standish to visit.

It's nothing great and fun to be here 24 hours, but it is convenient for emergencies. People say ghosts and this and that, but I never get a strange feeling in here. People say, Aren't ya afraid of dead people? I say, I'm afraid of live people. Live people can hurt ya.

We've had threats come in. Some people say: You're playing God. I say, you play God anytime you take a cancer treatment. I don't think it's playing God. I think it's just an extension of medicine. These people who die on the operating table for two minutes, drown for an hour, or freeze for a thousand years, as far as God's concerned there is no time limit. It's eternity.

When asked why he wanted to be frozen, Andy said only, *I don't wanna die.* As for the kind of future he expects to find after emerging from his long sleep, all he could say was that he hoped it would be good.

I remember when I first learned about death, Andy said. *We had a collie and she got hit by the school bus - slid on the ice and it hit her and killed her. So we're crying. I was in kindergarten. My brother was in first grade. It was a long, long time ago. What I remember was, dad had her on a sled. We just thought: stand her up, she'll walk, stand her up, she'll walk. She was laying on her side, dead, but there was no external... no blood out of the nose or mouth. She wasn't squashed. We just could not convince dad to stand her up and she'd walk. And he wouldn't do it! Stand her and hold her! But we were just convinced. I just couldn't get a grasp on that. We was convinced she was still alive or would be alive if you just stood her up.*

LP1

Being a pedophile is like living with a mask on, Shin Takagi confided in the midst of a crowded Tokyo cafe. The 47 year-old paused to light another cigarette. His mask was off today and people were noticing. In a sea of black business suits, Takagi sported a red Hawaiian-print shirt – daring them to look.

We should accept that there is no way to change someone's fetishes, Takagi insisted, jabbing his cigarette. *I am helping people express their desires, legally and ethically. It's not worth living if you have to live with repressed desire.*

Struggling to reconcile an attraction to children with a conviction that they should be protected, Takagi founded Trottla, a company that produces life-like child sex dolls. For over a decade, Trottla has shipped anatomically-correct imitations of girls as young as five to clients around the world.

I often receive letters from buyers, Takagi noted. *They say, 'Thanks to your dolls, I can keep from committing a crime.' I hear statements like that from doctors, prep school teachers – even celebrities.*

While our meeting that day was brief, Takagi invited me to visit his mountain workshop the following afternoon. I met him, along with my translator Natsuko at the Hachioji train station, an hour north of Tokyo. Our host was wearing a new Hawaiian shirt - bright yellow this time. On the twenty minute car ride over twisting forest roads, we talked about his own childhood.

It was a lonely time, the only son of a barber recalled. *I was always surrounded by adults. I would stay in my room and make paper crafts.*

I still live with my parents, the bachelor later confessed. *My father has Alzheimer's - but even before that, we never talked much.*

The solitude of Takagi's life is shared by his clients. *Most of them are men living alone*, he observed. *The system of marriage is no longer working. While most people buy dolls for sexual reasons, that soon changes for many of them. They start to brush the doll's hair or change her clothing. Female clients buy the dolls to remind them of their past, or to reimagine an unfortunate childhood. Many of them begin to think of the dolls as their daughters. That's why I never allow myself to be photographed. I want to prevent them from seeing me as the father of the dolls.*

The Trottla factory stands at the end of a remote gravel road, shrouded by trees. The building's only neighbors are monkeys, birds, and wild boar.

We had to be out in the wilderness, our host explained. *The machinery is loud and the materials flammable.*

Inside the dark interior, the stench of solvents was overpowering. Takagi admitted that the propriety solution he uses to replicate skin is a known carcinogen with toxic effects on the brain, liver, and kidneys.

It is a very challenging environment, he acknowledged. *That's why all my employees are former military. They are only allowed to work with the poisonous material two days a week and must always wear a mask and gloves. I often wonder what will kill me first - cigarettes or this.*

He hit a switch. The overhead florescent lights flickered on, and suddenly we were not alone.

At the far end of the room, hanging naked from metal stands were the dolls.

When I look at them in the middle of night, sometimes even I am scared, he admitted.

Does she have a name? I asked gesturing towards the nearest doll - a model he later described as a 10 to 12 year-old.

There is no name, he replied, *just a code name - LP1.*

What emotions do you see in her face? I pressed.

This one looks like she's sad, he responded. *One must make a variety of expressions to fulfil a variety of client needs.*

Takagi vacillated between objectifying and personifying his creations.

My objective is to create something that makes you say the doll is alive, he added. *My ultimate goal is to make a human itself. Only respect for God is stopping me.*

In Japan, with its animist Shinto beliefs, the dolls have a complicated status. *In Shinto,* Takagi elaborated, *everything has a soul. Even if you don't want the dolls anymore, you can't abandon them. There is a special ceremony that is performed for them at a shrine. It's like* a ceremony for a dead person. Since dolls have a human form, they must be treated as such.

He described a recent case in which a client who needed to get rid of a doll called, requesting his help. *He wanted me to dispose of it,* Takagi remembered. *But, he didn't say dispose. The phrase he used was send back home.*

At the end of our interview, as I was photographing a set of fiberglass molds, I noticed Takagi and my translator speaking in a corner.

What were you talking about? I asked her later.

My husband died in a motorcycle crash several years ago, she confided. *I was asking Mr. Takagi how much it would cost to make a replica of him.*

We all walked out together, the same way we came in - passing a pile of discarded doll skeletons waiting to be removed. These were the remains of some of the dolls that had been sent back home. *They're toxic,* Takagi explained. *So, we need a special company to come and pick them up. They have to be crushed with hammers.*
Everything with form must be broken eventually, he added.

On the drive back to the station, I asked Takagi if his work had changed the way he regarded the real and the artificial.

The manufacturer thought a moment. *It is a common belief in Japan that dolls are mirrors,* he said. *The dolls show their owner's true self.*

Code

Every particle in the universe is accounted for. The precise shape and position of every blade of grass on every planet has been calculated. Every snowflake, every raindrop has been numbered. Mountains rise sharply and erode into gently rolling hills, before subsiding into desert. Millions of years pass in an instant.

Here, in a dim room half an hour south of London, a tribe of programmers sit bowed at their computers, busy creating the universe - a digital cosmos as vast as our own. More accurately, through the science of procedural generation, they are making a program that enables the universe to create itself.

The project will be released as a video game under the title No Man's Sky. In the game, astronauts isolated from one another by millions of lightyears must find their own existential purpose as they traverse a galaxy of 18 quintillion stars.

The physics of every other game - it's faked, chief architect Sean Murray explained. *When you're on a planet, you're surrounded by a skybox – a cube that someone has painted stars or clouds onto. If there is a day to night cycle, it happens because they are slowly transitioning between a series of different boxes.* The skybox is also a barrier beyond which the player can never pass. The stars are merely points of light. In No Man's Sky however, every star is a place that you can go.

With us, Murray continued, *when you're on a planet, you can see as far as the curvature of that planet. If you walked for years, you could walk all the way around it, arriving back exactly where you started. Our day to night cycle is happening because the planet is rotating on its axis as it spins around the sun. There is real physics to that. We have people that will fly down from a space station onto a planet and when they fly back up, the station isn't there anymore; the planet has rotated. People have filed that as a bug.*

On the monitor before us, cryptic fragments of source code flash by - the laws of nature for an entire cosmos in 600,000 lines.

The universe begins with a single input, an arbitrary numerical seed - the phone number of one of the programmers. That number is mathematically mutated into more seeds by a cascading series of algorithms - a computerized pseudo-randomness generator. Machines, of course, are incapable of true randomness, so the numbers produced appear random only because the processes that create them are too complex for the human mind to comprehend.

Once the first seed number is entered into the void, the universe is unalterably established - every star, planet, and organism. The past, present, and future are fixed indelibly, with change to the system only possible from a force outside the system itself.

In one sense, because of the game's procedural design, the entire universe exists at the moment of its creation. In another sense, because the game only renders a player's immediate surroundings, nothing exists unless there is a human there to witness it.

Whatever is around you, Murray mused, *it actually doesn't matter whether it exists or not, because even the things you don't see are still going about their business. Creatures on a distant planet that nobody has ever visited are drinking from a watering hole or falling asleep because they're following a formula that determines where they go and what they do; we just don't run the formula for a place until we get there.*

The creatures, generated through the procedural distortion of archetypes, are each given their own unique behavioral profiles. *There is a list of objects that animals are aware of,* Artificial Intelligence programmer Charlie Tangora explained. *Certain animals have an affinity for some objects over others which is part of giving them personality and individual style. They have friends and best friends too. It's just a label on a bit of code – but another creature of the same type nearby is potentially their friend. They ask their friends telepathically where they're going so they can coordinate.*

While the basic behaviors themselves are simple, the interactions can be impressively complex. Artistic director Grant Duncan recalled roaming an alien planet once shooting at birds out of boredom. *I hit one and it fell into the ocean,* he recalled. *It was floating there on the waves when suddenly, a shark came up and ate it. The first time it happened, it totally blew me away.*

The formulas that govern behavior themselves are interdependent, yet still parametric. Murray evoked the simplified example of a sine curve graphed with linear time on the x-axis. Rather than simulate an entire creature's lifespan to determine its current status, the program need only enter the current x-value to find its y-axis behavioral counterpart. Due to computer processing limitations, every day is an exact reenactment of the day before – until a player arrives to alter the delicate equilibrium.

Similar processing limitations or creative preferences have led the team to separate many systems within the game. It was purely for aesthetics, for instance, that Duncan insisted on permitting moons to orbit closer to their planets than Newtonian physics would allow. When he desired the possibility of green skies, the team had to redesign the periodic table to create atmospheric particles that would diffract light at just the right wavelength.

Because it's a simulation, Murray stated, *there's so much you can do. You can break the speed of light - no problem. Speed is just a number. Gravity and its effects are just numbers. It's our universe, so we get to be gods in a sense.*

Even gods though, have their limitations. The game's interconnectivity means that every action has a consequence. Minor adjustments to the source code can cause mountains to unexpectedly turn into lakes, species to mutate, or objects to lose the property of collision and plummet to the center of a planet. *Something as simple as altering the color of a creature,*

Murray noted, *can cause the water level to rise.*

As in nature itself, the same formulas emerge again and again - often in disparate places. *If I were to take a normal cloud,* Murray reflected, *and imagine it made out of rock and brought down to earth, it would look a lot like a mountain.*

Particularly prolific throughout No Man's Sky, is the use of fractal geometry - repeating patterns that manifest similarly at every level of magnification. *If you look at a leaf very closely,* Murray illustrated, *there is a main stock running through the center with little tributaries radiating out. Farther away, you'll see a similar pattern in the branches of the trees. You'll see it if you look at the landscape, as streams feed into larger rivers. And, farther still – there are similar patterns in a galaxy.*

When I go out in nature, I don't even see terrain anymore, the programmer laughed. *All I see are mathematical functions and graphs. I'll pick up a stone and begin thinking about the shape of it. What formula could have given you that?*

The programmer reported recurring dreams in which the real world appeared to be just a computer program. The programmer considered the probabilities before offering a hedge. *Even if it is a simulation,* he said, *it's a good simulation, so we shouldn't question it. I'm working on my dream game, for instance. I'm more happy than I am sad. Whoever is running the simulation must be smarter than I am, and since they've created a nice one, then presumably they are benevolent and want good things for me.*

As a creator yourself, I posed, *how benevolent are you?*

Well, we don't have blood in our universe. That's pretty nice. We don't have cities full of urban problems. We have nice beautiful landscapes more often than not.

In No Man's Sky, there is also no sickness, no excrement, and no birth. There is death, but always with the assurance of reincarnation. *When you die, you regenerate in the same location,* Murray explained, *but you do lose a great deal of things. We wanted the loss to be meaningful - for you to know that if you make a decision, it has significance.*

The poignancy of death extends to other creatures as well. *The nature of video games is conflict*, Murray insisted. *It's an interesting reflection of where we've gotten to. With our game though, you give someone a controller - they land on a planet - they see an alien creature, and if it's their first time playing, they will probably shoot it even though they have just gone through a journey to get there.*

What I really like though, is that nine times out of ten, people suddenly feel bad that they've done it. You don't get points for killing. There are no gold coins. You chose to do that.

The player has no alter ego to hide behind either. *In most games, you begin by choosing a character*, Murray described. *Often you'll be cast as an unlikable character with a dozen catchphrases. You'll have a nickname like Irish or Tex. You're made to decide at the beginning who you are, but that might be before you decide how you really want to play. We want to let people have their imagination. They can be whoever they want to be. They might be an alien if that's what they want to believe. I quite like that.*

In a universe designed without mirrors, as this one is, the only way that you could ever view yourself would be to ask another player to look at you and describe what they see. Considering the inconceivable vastness of the cosmos however, for two humans to ever chance upon one another would be an almost impossible event - one capable of evoking real awe.

For the No Man's Sky team, that feeling of awe is the objective. In the words of programmer Hazel McKendrick, *You're not the god of this universe. You're not all powerful. You can't build a gun so big that you're unstoppable. You should be small and a little bit scared, I think, all the time.*

A sense of ecstatic obliteration once permeated Murray's childhood deep in the Australian outback. *My parents managed this big ranch of one and a quarter million acres,* he recalled. *It had a gold mine. It had seven airstrips. You don't get there by road - you have to fly in. We were very much on our own, and we went out every morning to check that the machines that were keeping us alive were still working. It was the closest thing to the surface of Mars. We were alone for hundreds and hundreds of miles. There was just this incredible feeling - knowing that you're this little dot in this massive landscape.*

The very first thing we talked about when we were planning this game was emotion, Murray continued. *That emotion of landing on a planet and knowing that no one else has ever been there before. There is a very deep human need to explore. When other games have exploration, everything has already been built by someone. There is a vocabulary. Certain doors will open and certain doors won't, and when the door opens, it probably has a little secret inside - a secret shared by thousands of other players that have been there before.*

Through the use of procedural generation, No Man's Sky ensures that each planet will be a surprise - even to the programmers. Every creature, alien spacecraft, or landscape is a pseudo-random product of the computer program itself. The universe is essentially as unknown to the people who made it as it is to the people who play in it - and ultimately, it is destined to remain that way.

People will stop playing long before even .1% of everything has been discovered, Murray reflected. *That's just how games are. I would be foolish to think anything else. It's a sad thought though. When we fly through the galactic map, we see all the stars, each of which will have planets around them, and life, and ecology - and the vast, vast, vast majority will never be visited. At some point the servers will be shut down. It will all be turned off, and it will be us who pull the plug.*

Book V

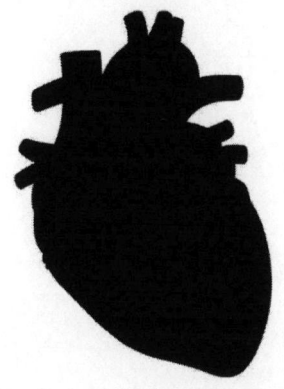

Sigrid

Sigrid appeared, pushing a pram down Mariannenstraße a decade later - her breasts swollen with milk. Blood rushed to my lips and cooled there.

She laughed that banshee laugh - as if she'd tousled my hair and broken all my combs.

We had first met on a Greyhound bus somewhere in the Midwest.

I asked you for the time, I recalled. *I asked you for the time and didn't listen to the answer.*

No, she insisted. *You asked me for a cigarette.*

The boy in the pram began to moan. She hoisted him, brusquely pulling up a corner of her blouse and tugging back her bra. His lips puckered out. He twisted and kicked joyfully - an enormous leech.

Back then, we had jumped off the bus in some dying town. The switchgrass sighed and sucked light as we walked on through empty lots under sagged telephone wires. The wires hung decaying shoes like Christmasy baubles. Looking down, even the bricks had given up and reversed into fistfuls of sand.

We crossed a moat of gravel, great-horned beetles, and barbed wire. We found no flowers, only flowerbuds - and, unsatisfied, tore open their seals to see future flowers.

All around us were little growth-stunted horses. They watched us with eyes like film-less cameras - frozen, as if they had forgotten their ability to fly. And this one white horse with its backwards legs and swollen drum-belly, its misplaced tail and forgotten teeth, its crowded heart, its black heat.

You said: Let's name it. Like when the world began. And then, we'll name ourselves.

You said: We should have many names. We change so much. We should have different names depending on whether we're happy or sad.

Different names in different stages of undress...

Or whether it's spring...

Or whether it's raining...

We won't always be together, I said sadly.

But, I am here now, she replied.

Yes, I said, *we are together now.*

And then the future came - making liars of us all.

Inbound Letter

You walk around Jerusalem and see all these old buildings built from special stone called Jerusalem Stone. Can you imagine all these white blocks of stone furrowed like thinking brows?

Well, a lot of Israeli developers wanted to knock those buildings down and construct modern high-rises, but there was an uproar from preservationists. So, what the government decided as a compromise was that the old buildings could not be destroyed. Instead, they must be moved to another location.

Now, the builders thought that all Jerusalem stone was the same. It didn't matter how they moved the stones. But, the preservationists *said, no, that is unacceptable. They have to be moved and rebuilt exactly as they were.*

So, what they do is, they number every stone on the building. You drive around Jerusalem and you see these buildings that are covered in writing. Literally every stone is numbered in order, so that a building can be exactly replicated - every stone, precisely where it was.

But, no matter how careful you are, it's not really the same building when it's in another place, is it? And, isn't that how memory is: something that you've marked, dismantled, moved, and reassembled somewhere that it doesn't belong?

Outbound Letter

Before you vanished, your edges blurred - and by then I was too preoccupied to pull the curtain back from the window.

That morning, at the hotel, I stared at a picture of your face from some annihilated day. Luke was beside you then as he is beside no one now. I thought of the billion-year-old rock on which he sat at the bottom of that gorge as he wept for his own existence - and felt structures straining and snapping all throughout the earth.

I have inhabited photographs of you. To look at them is to wear them as masks. I made those images gaze at me, directly in my eyes. When I breathed, the photographs swelled with life.

I sense your shape beyond the things I have seen. Inside of this sense, fits the world. The farther you are, the vaster the world.

This separation is our rehearsal for dying.

Among your photographs, were those I had bought from a flea market. They had been ripped from family albums of people I had never met.

Beside the tomb, the girl and the camera shutter blink in unison, forever. The man chases a woman crookedly through crooked pines. The soldiers shave each other's cheeks. The infant nurses. The black dog speaks.

I intercept the glances of lovers.

What will become of the trapezoid, you asked, made by your chin, my arm, your neck, my breast?

In this clutch of snapshots, the groom's face has been savagely cut out.

The radiant bride stands at an altar smiling at a hole, feeds cake to a hole, speaks her vows into a jagged little hole containing everything that has ever been lost.

Alexandra

The further away I go, you wrote, *the closer I'll be to you.*

The day we met, you sat on the grass, becoming smaller and smaller as I climbed a tree - branch by branch, to be closer to you.

You wrote, *How long until we see each other again?*

You wrote about measuring distances by the degree to which we can comprehend them.

You died on a Wednesday.

I saw your man at the funeral. *I remember you*, he said, although we had never met before. We wept, holding one another. His hand slipped into mine with a lover's familiarity. Vicarious knowing.

She is the grass, the cloud, the table, he said.

That breeze, you wrote, *that tickled your ear? That was me.*

Your mother believes that you have become an angel, capable of intervening in this world.

The last time I saw you was in a dream. Red hair flying, apple-devouring mouth, pupils dilated with love. You were waiting on the far shore, but the current swept me away. I knew that I was dreaming, and warned all the people I passed, but they refused to believe. You, me, and everyone else in that place - lonely orphans of a single dreaming mind.

A Children's Story

*A*mphibian imps known as Kappa live in the lakes and rivers of Japan, Rin recited. *At the crown of each of their heads is a depression filled with water, their life-force. Kappa are fond of eating children, yet are also polite. Therefore, if a child sees a Kappa, the child is taught to bow. The creature, forced by etiquette, must bow as well. The water spills. The creature dies.*

Ukiyo

I lived for a while in a small tatami room in the Asakusa district of Tokyo - once the greatest pleasure quarters in all of Japan. Its brothels, tea houses, and Kabuki theaters were known as Uyiko: The Floating World. Samurai abandoned their swords at the gates.

In 1945, the area had been leveled by American firebombing. The B-29s flew in at 500 feet, close enough to smell burning flesh - to suck up scraps of singed clothing into their fuselages.

Asakusa was reincarnated as a quiet temple district, although periodically someone digging in her garden or renovating his kitchen will unearth an unexploded firebomb - a metal tube the size of a forearm designed to spray flaming napalm 100 feet.

Had we been alive at the time, I might have flown in the Air Force and turned the girl lying next to me, Rin, into one of the: *swollen, contorted, blackened bodies that resembled enormous ginseng roots*, as a witness described.

And when my plane crashed in a red mist, Rin would have plunged her bamboo spear into my belly, up into the surge of my fluttering heart. *With love*, she promised. Just another way to know each other.

But now, origamied into one another, we listened to a truck outside trolling the roads. Its repeating megaphone announced: *It is getting late now. All good children must go home.*

Rin did not consider herself one of the good children. She was seventeen, in her school uniform, a skirt and sailor blouse that must be worn even in winter.

The principal explained that cold builds character and the children should learn not to feel it. Rin stated that she had learned that lesson too well. Learning not to feel the cold was just the first step of learning not to feel anything at all.

She passed time in her room or wandering Tokyo until her boredom gave her such a look of drunkenness that perverse strangers would try to trick her, claiming, *I'm your friend. We were supposed to meet here. Let's go.*

She came over after school, bringing for me the lunch her mother had packed for her, untouched. Today's lunch consisted of hermetically-sealed waffles, rice with fermented plum, eggplant in sauce, tuna with radish, and a thermos of tea mixed with chocolate.

I said, *You're good to me.*

She said, *It amuses me.*

It amuses you to be kind?

It amuses me to do something. Not the same as kindness.

Underwater

Rin and I met in Ueno Park. I met many girls there - evolving a certain routine. Always as we descended the stairs I would repeat what an elderly man from the tourist bureau had told me: *10,000 years ago, all of this was underwater.* I would add, *Can you imagine the fish swimming in the air all around us?* My hand would become a fish and the fish would bite her ear and she would giggle.

We would cross the little bridge to a Shinto shrine, a great slab of black polished stone carved with an image of eyeglasses. I would explain that this was a shrine to eyeglasses. Worshippers would thank glasses for allowing them to see. There was another shrine to honor catfish killed by fishermen and another for kitchen knives.

Then I would peek into the keyhole of a temple and pretend to catch a glimpse of God.

But, Rin hated the park. She hated the dark. She hated the cold. She hated the lake with the wilting lotus plants. She hated the shape of the temple, the stamens of flowers, the dogs, the hardness of rocks, the taste of the air.

I said, *It's fine to hate things. Hating things is like loving them. You are noticing them. You lavish them with attention. As long as you aren't indifferent.*

No, she said, *I am never that.*

Rin also hated her body, her shoulders so broad and strong. Kids joked about her being a football star. Her legs so thin and hungry-looking. Seeing her naked for the first time, lying on my mat, I imagined she was dying, withering from one end to the other. Our naked bodies were covered in the sweat that looks like glass. Skin was no longer the surface of the body, and the hider of things, but rather an inner layer, exposed as if in a display case.

I said, *You look like a virgin.*

Well, she responded, as if blowing out a candle, *I'm a virgin to you anyway.*

Rin's First Letter

*Y*ou should not have had sex with such a stupid person like me...
I am the most stupid person in this world...

My problem is that I easily have sex with a person whom I do not know very well...
I know this is not a right thing to do... and I don't do this because I like sex.

Most of the time when I have sex with someone that I do not know well, I do not enjoy doing that. But I cannot refuse it because that is the only moment I feel that I am not alone...
I am kind of addicted to it...

If I do not have sex for a long time, one day I cry like a monster and cannot stop...

To stop crying, I take my older sister's pills. I

do not know what they are but they make me very sleepy and I sleep for almost two days. I become emotionless...
Perhaps I am crazy... I wanna be normal...

You don't have to reply to this, I just wanted tell you how stupid I am...

Good night, and again I am so sorry for everything.

Rin's Second Letter

I wish I knew how to control my emotions like you do...

A long time ago, I decided not to cry in front of others...

When my older sister started having eating disorders I was really sad but didn't want my parents and my sister to worry about me.

Then I found a place to cry, which was bathroom. I used to cry a lot there while I was taking a bath.

Then I met you and you let me cry as long as I wanted to... and that was really nice of you.

I always wanted someone who could be next to me when I'm crying.

Sorry if I made you feel bad sometimes because I cried a lot. That wasn't your fault at all and there was nothing you could do. You gave me a new space where I could cry and that was enough. Thank you a lot.

Now you leave to USA and I'll go back to my bathroom when I feel like crying...

I don't know what I wanted to say but thank you very much for everything. Take care and enjoy your life.

Rin's Third Letter

It is not easy to fully accept this...
In some parts of my mind, I still cannot believe this...
But sometimes I feel very happy about this...

Even though it is still bit cold here, I feel something warm in my body.
It is a strange feeling... a feeling that I haven't had in my entire life...

I am happy because I am not alone.

Even though I am alone in my room, in my house, or wherever...
There's someone else inside of my own body, and he (maybe she) will never leave me alone till I abandon him.

Perhaps this is what I wanted for a long time, neh...
But I am not ready to raise this baby...
I am not even ready to listen to what my parents and sisters say about this baby...
Not ready to face the difficult things...
Not ready to give up the opportunities that I have, even though I don't know what I want to do with my life...

I am pretty sure that I am going to have an abortion...
But I will never forget about this baby.

I have decided his birthday. His birthday will be November 15th, 2007. Every year, I will celebrate this day, imagining my first child getting older, bigger... (You can join if you want to.)

There are services at temples and shrines here to hold a mass for unborn children.
It will cost me about 20000 yen.
I'm not sure if I am going to do this.
But if I've got extra money, I am going to do this.

You should visit him someday in the future.

Last night I had a strange dream.
I was in the hospital and had an abortion.

I had to go to the doctor's office again, and the doctor told me that everything was done, and he added one sentence when I was about to leave his office. He said, It was a beautiful 3 months old girl.

In my dream I was shocked to know that it was a girl because I had been thinking that it was a boy for all the time.
I was also sad to know that I had lost my baby.
I realized that I will never be able to see her...
And then, I woke up.

I thought what the doctor said in the dream was strange...

He cannot say that the baby was 3 months old if it is not born yet...
And how could he know that she was beautiful,
if her face is not fully formed yet...?
It was a strange dream...

I have had a lot of dreams about this.
Sometimes in dreams I was trying to justify having an abortion... and other times I was blaming myself on this...

Rin's Fourth Letter

I met another boy...

He is kind of stupid.
He is very stupid.
He is very very stupid.

But nice, neh.

I had sex with him just one time after you left.

When I found out I was pregnant, I said that the baby is his and it was not true...

Why? Because I was so angry and I didn't have someone to be angry to. So I was angry to him, punching him, kicking him...

And it's your fault because you were not there.

But he was so stupid...
He did not do the calculations to see that he could not be the father...

Maybe not stupid.
Maybe innocent...

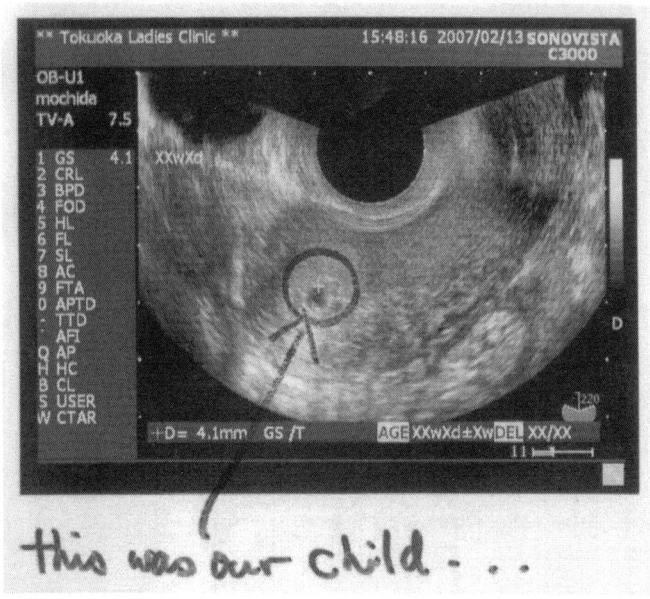

this was our child...

Love

Crawling back into the ambulance, onto the stretcher - I slept, and had this dream:

I was in a field. An orb loomed overhead - silvery and flickering like a beetle's wing - just blowing in the breeze. A woman carried me inside. She stripped me bare, stripped away my clothing. Insatiable, she peeled away my skin. With brailleing fingers, she devised my seams. She popped and reeled my nerves in tangled coils. She plucked ripe organs from my chest - collapsed the rest. Chewed my bones and blew the dust.

I fell in love. It was one of those dreams where the dreamer loves someone who does not exist. The dreamer has fallen in love, and when he awakens, he is heartbroken - heartbroken with the missing of himself.

Printed in Great Britain
by Amazon